Treasures

of The Royal Horticultural Society

Brent Elliott

The Herbert Press
in association with The Royal Horticultural Society

Copyright in the text © 1994 The Herbert Press Ltd
Copyright in the illustrations © 1994 The Royal Horticultural Society
Copyright under the Berne Convention

First published in Great Britain 1994 by
The Herbert Press Ltd, 46 Northchurch Road, London N1 4EJ
in association with
The Royal Horticultural Society, London

House editor: Julia MacKenzie
Designed by Pauline Harrison

Set in Garamond
by Nene Phototypesetters Ltd, Northampton
Printed and bound in Hong Kong by South China Printing Co. (1988) Ltd

A CIP catalogue record for this book is available from the British Library.

ISBN: 1-871569-68-0

Foreword

THE ROYAL HORTICULTURAL SOCIETY owns a remarkable collection of drawings, which has been built up since the Society started in 1804. The collection is still growing as works continue to be acquired, commissioned and given. The scope and nature of the drawings combines a huge range of plant types and a fascinating variety of drawing technique and skill.

The collection has not been widely known nor seen in the past; drawings are fragile and cannot be on permanent exhibition nor handled frequently.

In 1992, the Society's Picture Committee put forward plans to show a series of exhibits at our Westminster Shows during the year 1992–93, culminating in a major exhibition 'Treasures of the Royal Horticultural Society' which was held at the Kew Gardens Gallery in the spring of 1993.

In order to bring some knowledge of the Society's collections to a wide audience, a selection of paintings and drawings has been brought together to form the contents of this book. Seventy plates, arranged chronologically from the 1630s to the present day, cover a wide range of plants, from British wild flowers to glasshouse exotics. The introduction puts the drawings into the multiple contexts of art techniques, botanical nomenclature and the Royal Horticultural Society's history.

This is a start. I hope over the next four years that the Society will be able to publicize more drawings from the collection.

SIR SIMON HORNBY
President
The Royal Horticultural Society

Introduction

ON 7 MARCH 1804 a group of seven men met at Mr Hatchard's bookshop in Piccadilly to discuss the founding of a society 'to collect every information respecting the culture and treatment of all plants and trees, as well culinary as ornamental'. The seven were: John Wedgwood, of the famous pottery family, who had formulated the idea of the society in 1801; Sir Joseph Banks, the President of the Royal Society; William Forsyth, the royal gardener at Kensington; William Aiton, the royal gardener at Kew; the nurseryman James Dickson; and two amateur botanists and gardeners, Richard Anthony Salisbury and the Hon. Charles Greville. The result of their deliberations was the Horticultural Society of London.

The meetings of this new Society provided a forum for discussions about the improvement of horticultural practice and for the exhibition of plants. In 1821, the Society leased thirty-three acres of the Duke of Devonshire's estate at Chiswick to serve as an experimental garden; here it began conducting examinations for gardeners, and pioneered the development of the modern flower show. From the 1820s to the 1840s, the Society sent out a distinguished series of collectors, such as David Douglas and Robert Fortune, to introduce plants from abroad; these plants, identified at Chiswick by the Society's botanists, and distributed to members, ranged from the Douglas fir to Chinese garden varieties of chrysanthemum (see plate 38).

Despite all these successes, the 1850s were a time of financial crisis for the Society, leading to the sale of its herbarium and library. In order to revive its fortunes, its then President, Prince Albert, arranged a new charter under the title of The Royal Horticultural Society in 1861, and secured for it the lease of a site for a new garden in Kensington. This remained the Society's headquarters until 1888; the site has since been built upon, and the Science Museum and Imperial College now stand where the canals and fountains of the garden once were. (The Albert Memorial, the Albert Hall, and the principal buildings to the south are still aligned on the axis of the vanished garden.)

In 1903, the garden at Wisley, Surrey, of the industrial chemist George Fergusson Wilson, a former Council member, came onto the market; it was purchased by Sir Thomas Hanbury and presented to the Society to serve as a new experimental garden, away from the air pollution of London. The Chiswick site was relinquished, and has since been built over. In 1904, the Society's centenary year, it also acquired a permanent exhibition hall and offices on Vincent Square; in 1928 an additional exhibition hall was built around the corner on Greycoat Street. These buildings are still the Society's headquarters, and serve as the venue for most of its flower shows. Its great spring show in late May, after the move from Kensington in 1888, was staged at the Temple Gardens; in 1913 this show was moved to the grounds of the Royal Hospital in Chelsea, where (popularly known as the Chelsea Flower Show) it has remained ever since.

Throughout its history, the Society has been a leading force in the development of flower shows, of horticultural education, of plant registration, and of the application of scientific techniques to gardening. In 1858 its Fruit and Vegetable Committee, and in 1859 its Floral Committee, were established; these have since been augmented by a variety of other committees with narrower remits (Orchid, Narcissus and Tulip, and a number of Joint Committees set up with specialist societies). Part of the purpose of these committees was to give awards to plants, instead of simply to exhibitors, as had previously been the case. Two classes of award were instituted in 1858: the First Class Certificate, and the Second Class Certificate (replaced by the Award of Merit in 1888).

In recent years, the Society has extended its services throughout the country by developing regional centres, either in gardens that have been presented to it – Rosemoor in Devon and Hyde Hall in Essex – or in association with established gardens, such as Harlow Carr in Yorkshire, Ness in Cheshire, and Pershore College of Horticulture in Worcestershire. It has also set up two new organizations under its auspices: the Institute of Horticulture, as a professional body for gardeners, and the National Council for the Conservation of Plants and Gardens, which has established a network of national collections for different categories of cultivated plants, in order to ensure the survival of as large a range of garden plants as possible.

The Society has also developed what is probably the foremost horticultural library in the world, divided between a scientific library for staff and students at Wisley, and its principal and historical collection housed at Vincent Square. At the time of writing, the latter comprised some 47,000 volumes, over 1500 sets of periodicals (some 350 of them current), the country's largest collection of horticultural trade catalogues, and about 18,000 drawings. It is these last that this book is concerned with.

The Society's early artists

THE SOCIETY's first involvement with botanical art arose from its work with fruit. Thomas Andrew Knight, the Society's second President, alarmed by reports of the decline in quality of certain long-established fruit varieties, had advanced the theory that varieties, like individuals, had definite life-spans; it was accordingly a matter of some urgency to start breeding new varieties to replace those that would soon die out. Knight's theory was incorrect, but gave a great incentive to experimental fruit breeding.

The Society set up a collection of fruits in its garden at Chiswick, and found itself faced by the problem of nomenclature. The same fruit might pass under different names in different nurseries; there was great uncertainty over which names were synonyms. In 1824 Robert Thompson (1798–1869) was appointed to the fruit department, and worked there for the next forty-four years; in 1826 he published the first catalogue of the collection. There were then 3825 named varieties in the garden, plus another thousand whose names were less certain, and over the succeeding years, comparison of the fruits enabled their synonymy to be progressively determined. But publishing synonyms in a list was one thing; it was another to provide a clear and unmistakable type for each correct name.

In 1815, William Hooker was hired to paint twenty-five varieties of fruits each year, and the Society soon rejoiced in the collection it was building up, 'the importance of which as standards of reference, will long be felt and acknowledged'. After 1820, ill health slowed Hooker's production, and he stopped painting after 1822, probably as the result of a stroke. From 1820, the Society continued the fruit series by commissioning other artists: Charles John Robertson, Barbara Cotton, and Augusta Innes Withers. Examples of the work of these four artists can be seen in plates 27–30; in 1989, the Herbert Press published ninety-six of Hooker's drawings, with commentary, under the title *Hooker's Finest Fruits*.

The year after the depiction of fruits began, the Society found a further use for its painter: illuminated autographs. Hooker prepared decorated vellum sheets to bear the signatures of two of the Society's royal patrons – the two Charlottes, Queen and Princess – to be followed over the next few years by the Prince Regent and various members of foreign royalty appointed honorary members. When Hooker became too ill to continue, Elizabeth Francillon carried on, followed by Mrs Withers. The custom of royal autographs lapsed after 1840, but was revived by Queen Victoria in 1861. During the present century, it has been largely confined to royal patrons. Later artists included W. H. Fitch, Frank Galsworthy, E. A. Bowles, Lilian Snelling, and the late Emily Sartain.

John Lindley

IT WAS NOT ONLY OF FRUITS that pictures were commissioned. Hooker, Mrs Withers, Miss S. A. Drake, and others portrayed various kinds of ornamental plants, some for publication in the Society's *Transactions*; other drawings arrived as gifts, either individually, as in the case of James Sillett's dahlia portrait (plate 33), or as a series, like the drawings of Chinese plants sent by John Reeves. The work of identifying new plants, and of organizing their illustration, fell to John Lindley. Since, through his work for the Society and outside it, he had a great impact on botanical art, and since the Society's library is named in his honour, it is appropriate to say something about his life and work.

John Lindley was born in 1799, the son of George Lindley, a nurseryman at Catton, near Norwich, whose book on *The Orchard and Kitchen Garden* he later edited for publication. In 1818 he went to London, following in the footsteps of his slightly older friend, William Jackson Hooker (later Director of Kew), who introduced him to Sir Joseph Banks; Lindley spent eighteen months working in Banks's library, and, on Banks's death in 1820, was hired by William Cattley of Barnet to publish the *Collectanea Botanica*, a series of illustrations of plants from his collection.

In 1822, Lindley was appointed Assistant Secretary to the Horticultural Society at its garden at Chiswick, his duties being 'to have superintendence over the collection of plants, and all other matters in the Garden'. In 1827 he was elevated to the position of Assistant Secretary to the Society as a whole, with duties at both Chiswick and its London offices, and in 1841 was appointed Vice-Secretary.

His work for the Society did not prevent Lindley from accepting a range of duties elsewhere. In 1828 he was elected Professor of Botany at the newly founded University College, a post he was to hold for over thirty years in addition to his duties on behalf of the Horticultural Society. From 1835 to 1853 he was also Professor of Botany to the Society of Apothecaries, a role which brought with it the responsibility of being director of the Chelsea Physic Garden. In 1826 he became editor of the *Botanical Register*, which had been founded in 1815 by Sydenham T. Edwards; from 1828 to 1830 he edited the *Pomological Magazine*, and from 1846 to 1855 he edited the *Journal of the Horticultural Society*. In 1841, together with Joseph Paxton, he founded the *Gardeners' Chronicle*, of which he remained the horticultural editor until his death; the longest-running of horticultural periodicals, this still continues under the new title of *Horticulture Week*. Lindley and Paxton further collaborated on *Paxton's Flower Garden* (1850–53).

Lindley borrowed for his other magazines the artists who worked under his supervision at Chiswick. For the *Pomological Magazine*, he used the services of Augusta Innes Withers; for the *Botanical Register*, he used Miss Drake.

Nor does this exhaust the list of Lindley's extra-curricular activities. In 1838 he compiled a report on the condition of the royal gardens at Kew, which led to the establishment of the Royal Botanic Gardens, of which his old friend William Jackson Hooker was made the first Director. He was active on the 1845 commission to enquire into the causes of the Irish potato blight; he was a juror for food products at the Great Exhibition of 1851; and for many years he was consulted by the Admiralty about the planting of the island of Ascension. He was the first botanist to work out a classification of orchids, and wrote prolifically on the subject (the American Orchid Society has named its scientific journal *Lindleyana* in his honour). He also published monographs on roses and foxgloves, on fossil botany, and several works on plant classification.

In 1858 Lindley was at last promoted to the position of Secretary of the Society. His last major function was to act as one of the organizers of the Great Exhibition of 1862, a sequel to that of 1851, which was held in the Society's new garden. By the time the Exhibition was over, Lindley's ill health obliged him to relinquish his official positions. His last years were spent suffering from failing memory and 'softening of the brain'; he died on 31 October 1865. His estate, Bedford House, Acton, near the Society's garden, became the site of London's first garden suburb, Bedford Park, the course of whose streets was planned in order to preserve as many of his trees as possible.

Disaster, and a new start

DURING THE 1850s, the Society entered a period of financial decline, and the closing years of the decade saw the sale of its herbarium, a quantity of plants from its garden, and, in 1859, its library.

The story of the library had begun in 1806, when Dr John Sims presented five French gardening books to the Society. The following month, a copy of Philip Miller's *Gardeners Dictionary* was purchased, and by 1813 there were enough books to justify buying a bookcase. A Library Committee was established in 1817, and at its first meeting, it recommended the purchase of Redouté's *Les Roses*; it was plainly one of the Society's more expensive committees. Over the ensuing four decades the library amassed an impressive collection of books and periodicals.

The library was sold at Sotheby's, in a four-day sale, 2–5 May 1859; the total sale price was £354 2s. The sale did not include the royal autographs, but all the other drawings went. The Hooker fruit drawings sold for £49 10s, the Reeves Chinese drawings (see below) for £70. These at least have since come back to the Society by various routes, but other collections sold at the time are now scattered, and the current whereabouts of many is not known. Among these losses were passionflowers by Ferdinand Bauer (now in Cracow), waterlilies by Barbara Cotton, daffodils by John Curtis, chrysanthemums by Curtis and E.D.Smith, peonies and dahlias by Clara Maria Pope, designs for glasshouses by John Claudius Loudon, and the original drawings for the Society's medals and certificates.

The Society soon began to think of replacing the lost library. In 1865, an opportunity arose, with the death of John Lindley, whose library was put on the market in 1866. The Society had just helped to organize the first International Botanical Congress, and some £600 of the resulting profit was used to purchase Lindley's collection. In 1868, a group of Council and Committee members formed the Lindley Library Trust, purchased the collection from the Society, and invested it in this Trust, so that it would never again be put in danger of being sold. Maxwell T. Masters, the last of the original Trustees, died in 1907, and the Council of the Society was made the sole Trustee.

Some important groups of drawings formed part of the Lindley purchase, most notably Dean Herbert's crocus drawings and Ferdinand Bauer's illustrations (see plates 25–26) for Lindley's *Digitalium Monographia* of 1820 (including five drawings by Lindley himself). Lindley was a good amateur artist, as will be seen from one of his rose drawings, reproduced in plate 31.

The later growth of the collection

UNTIL THE 1930s, the drawings collection grew sporadically and depended mainly on gifts and bequests. In 1910 Henry Wyndham presented two volumes of drawings of Surinam plants made around 1830 by George Schouten and John Henry Lance (after whom Lindley had named *Oncidium lanceanum*). Among the few purchases to be made were four volumes of seventeenth- and eighteenth-century German drawings, originally collected or commissioned by Christian Wenzel von Nostitz-Rieneck (1649–1712), bought for the Society by Colonel Stephenson Clarke of Borde Hill in 1930. In 1926, the Hooker fruit drawings, sold in 1859, were purchased from John Napier of Leighton.

In 1936 the Library received its greatest single bequest: the library of Reginald Cory (1871–1934). Cory was the son of a Cardiff coal millionaire, and his garden at The Duffryn, today open to the public, became an important garden at the turn of the century, especially when, on the eve of World War I, he made it the venue for dahlia trials, helping to spur a revived interest in that flower. When the Royal Horticultural Society took over the publication of the foundering *Curtis's Botanical Magazine* in 1922, Cory financed the publication of the missing 1921 volume (which was not completed until 1938). Cory bought books and drawings like a magpie, accumulating them faster than he could unpack and sort them. His collection paid a price: when a fire broke out at The Duffryn, many of the books and volumes of drawings suffered water damage from the firemen's hoses. Nonetheless, in both quantity and quality, Cory's bequest furnished the core of the drawings collection. Almost all the Society's seventeenth- and eighteenth-century drawings came from Cory: the Kouwenhoorn florilegium (plates 1–2), the anonymous tulip drawings entitled 'Hortus Florum Imaginum' (plate 5), hundreds of ink drawings by Claude Aubriet, an album of drawings by Ehret and Taylor made for the Duchess of Portland (plate 10), further Ehret drawings made for the Earl of Bute, and an immense collection of unsigned drawings from Bute's library, most of them bound under the title of 'Flora Asiatica' (plates 12, 13).

Among the other trophies of the Cory bequest was an album of drawings on vellum by Turpin; the copy of his *Leçons de Flore* made for Louis XVIII (plate 35); and the miscellaneous portion of the Reeves Chinese drawings, which had been sold as part of the Society's Library in 1859.

No subsequent bequest has been as extensive as Cory's, but there have been others worth noting. E. A. Bowles (1867–1954) bequeathed not only many of his own drawings (plates 51, 52), but also several by the botanist and plant collector F. W. Burbidge – both the scrapbook of drawings he made while collecting in Borneo for the Veitch nurseries in the 1870s, and some drawings of daffodil bulbs incorporating fragments of bulb tunic pasted onto the paper. Gurney Wilson (1878–1957), the former editor of *The Orchid Review*, bequeathed a collection of orchid drawings by various hands (plate 70). Many individual artists in recent times have either bequeathed or presented quantities of their work, among them Vera Higgins, Lilian Snel-

A watercolour sketch of a *Solanum* species by Frederick William Burbidge (1847–1905), from the scrapbook he kept on his 1877–8 expedition to Borneo

ling, and Admiral Paul Furse, examples of whose drawings will be seen in this volume (plates 53–4, 59, 62).

The Society's collection also includes drawings deposited on permanent loan by their owners. These include the Thomas Edgerley drawings, made by a gardener at Tatton Park in the 1820s and 1830s; an album of drawings made on his African and South American travels by Sir Charles Bunbury, lent by his family (see plate 43); and drawings of irises by Paul Furse and Harold Round, lent by the British Iris Society.

The Society's illustrated publications

ANOTHER SOURCE through which the drawings collection has been augmented is the Society's own illustrated publications. For the Society's first periodical, the *Transactions* (1805–48), many drawings were commissioned that were published as hand-coloured engravings; these were sold in 1859, but some, most notably the fruit drawings, have since been recovered. For its later periodicals, the *Journal of the Horticultural Society* (1846–55), the *Proceedings* (1859–65), and the *Journal of the Royal Horticultural Society* (1866 to date, and retitled *The Garden* in 1976), it relied first on wood-engravings and later on photography, and no longer commissioned drawings. However, some ink drawings for particular illustrations have entered the collection, as have ink drawings for some articles in the Society's later yearbooks – most notably a group of drawings by W. A. Constable, used in his article 'The Comparison of Lily Bulbs' in the 1946 *Lily Year Book*. The collection also includes drawings commissioned for some of the books published by the Society, most notably George H. Johnstone's *Asiatic Magnolias in Cultivation* (1955), Sir Frederick Stern's *Snowdrops and Snowflakes* (1956), and the *Supplements to Elwes' Monograph on the Genus Lilium* (1960–62). In 1979 Hugh Johnson started a quarterly journal, *The Plantsman*, to complement *The Garden*, although published by New Perspectives Publishing Ltd rather than by the Society; some of the drawings commissioned for it have been added to the Library (see plate 67). In 1993, the Society started *The New Plantsman* to succeed the previous title, and a selection of the drawings commissioned for it will be acquired for the Library henceforward (see plate 69 for the first example).

The Society and botanical illustration

AT ITS THREE WINTER SHOWS each year, the Society holds exhibits of botanical art, at which artists, both amateur and professional, may display their work and compete for medals. To judge these exhibits, the Society has a Picture Committee, whose minute books begin in 1934. Among the members of the Picture Committee over the years have been Wilfred Blunt, the author of *The Art of Botanical Illustration*, E. A. Bowles, Sir David Bowes-Lyon, the botanist John Gilmour and the artist John Nash. The Committee awards the Grenfell Medal (instituted 1919 to commemorate Field-Marshall Lord Grenfell, the Society's President, and designed by E. A. Bowles) and the Gold Medal for exhibits of pictures or photographs. In addition, the Lindley Medal (instituted 1866) can be presented for illustrations of 'special scientific or educational interest'.

In 1985, at the suggestion of the former Librarian, Peter Stageman, it was determined to allow a budget for the purchase of drawings exhibited at the shows, and new drawings have been purchased each year since 1987.

Oriental drawings in the Society's collection

AN IMPORTANT PART of the Society's drawings collection consists of drawings by oriental artists. These, unfortunately, are in almost all cases anonymous.

The most important group is the collection of drawings of Chinese plants made by (unnamed) Chinese artists under the direction of John Reeves (1774–1856). Reeves was an inspector of tea at Canton and Macao for the East India Company, and in 1817 he offered to send plants and drawings to the Society. By 1819 the Society had received 130 drawings, and others continued to arrive during the 1820s, reaching a total of 755. Some of the drawings bear John Lindley's annotations; some of them in fact are the type specimens for botanical names which Lindley coined without having seen the original plant. At the sale in 1859 they were sold in two sets: one consisting of 127 drawings of camellias, chrysanthemums, and peonies, the other comprising the remaining 628 drawings. Together they fetched £70.

The miscellaneous Reeves drawings came back to the Society as part of the Cory bequest. The group of 127 drawings, bound in three volumes, was bought by the Society in 1953 from Messrs Heywood Hill for £400 – a sum nominally greater than the entire sale price of the Library in 1859. Two of these drawings are reproduced as plates 37 and 38.

In addition to the Reeves drawings, there are two other groups of Chinese drawings: a collection of 100 anonymous nineteenth-century drawings (see plate 39), and an album of fifty drawings, mostly of fruits and vegetables, made in Canton about 1800 and bearing the name of Wang Liu Chi, though whether as artist or owner is not certain. There are also two albums of Japanese drawings: one of these, illustrated in plates 40–41, depicts a mixture of well-known garden varieties, and exotic plants introduced into Japan for garden use; the other depicts lily cultivars in a cruder style. One of the items purchased in 1866 as part of John Lindley's library was an album of drawings by Harmanis De Alwis Seneviratne, draughtsman to the Peradeniya Botanic Garden in Sri Lanka in the 1820s, which includes the earliest known plan of

that garden. There are also some groups of drawings made by Indian artists.

Most of these drawings are not typical of the traditional floral art of their countries. Although made by oriental artists, they show a great debt (though not complete assimilation) to European artistic style, either because they were directly commissioned by British patrons, or because the artists were conscious of a western market. Western botanical art had already made an impact in Japan before its opening to the west in the 1850s, and by the turn of the century the Yokohama Nursery Company was printing trade catalogues with coloured plant portraits for the foreign market.

Garden scenes

THERE IS ALSO a small number of garden drawings in the Society's collection. The largest number of these consists of the plans made for the Kensington garden, including drawings by the landscape gardener William Andrews Nesfield, the architects Francis Fowke, Sydney Smirke, and Saxon Snell, and the sculptors Godfrey Sykes and Joseph Durham. In addition, there are garden scenes, usually in watercolour, which have been received by gift or bequest over the years: six mid-nineteenth-century views of the gardens at Elvaston Castle, Derbyshire, by George Maund; two Japanese scenes by Ella Du Cane; a group of garden sketches by E.A.Bowles; some two dozen garden scenes made in the 1920s by Edith Helena Adie, many of them depicting Reginald Cory's garden at The Duffryn; and a few sketches by the garden designer Ralph Hancock, most notably a large perspective drawing for his famous roof garden at Derry and Toms in Kensington (now British Home Stores).

Award portraits

THE LAST CATEGORY of drawings in the Society's collection consists of portraits of orchids that have been given awards at the Society's shows.

In 1897, Nellie Roberts was commissioned to paint portraits of orchids to which the Society had given awards; she continued to serve as the Society's orchid artist for over fifty years (occasionally, other artists, such as Stella Ross-Craig, stood in for her). She was awarded the Veitch Memorial Medal in 1953. Since her retirement, a series of distinguished artists has continued the work: Jeanne Holgate, M. I. Humphreys, Gillian Young, Jill Coombs, David Leigh, and Cherry-Anne Lavrih. The number of orchid portraits now stands at about 6000, constituting a third of the entire collection. The orchid drawings are separately housed, in order to be available for the use of the Orchid Committee.

For a brief period between the wars, the attempt to portray award-winning plants was extended beyond orchids, with Alfred J. Wise as the principal artist from 1926.

The purposes of botanical art

THE SOCIETY'S COLLECTION of paintings and drawings is plainly an eclectic one, arising partly from commissioned work and partly from the accidents of gift and bequest. Part of the collection serves botanical and horticultural purposes; another part acts as a record of changing styles in plant portraiture over the last few centuries.

In looking at the drawings reproduced in this volume, the reader should ask, what were the artist's purposes? and how were these purposes achieved?

Anyone studying botanical art today will quickly be instructed about the difference between botanical art and mere flower painting: historically, however, it is not always possible to make a clear distinction. Botanical art could not be said to exist until botanists had distinct requirements to be met. The primary purpose of botanical art is the identification of plants, but just what that entails depends on the system of classification that is being used, and what factors are held to differentiate plants.

The earliest printed herbals carried illustrations that were copied from medieval manuscripts, and which frequently bore little resemblance to the intended plants. From the publication of Brunfels' *Herbarium Vivae Eicones*

(1530) on, the artists of the early herbals set themselves the goal of depicting plants so that they could be identified in the field; and until the requirements of botanists became more exacting, simple visual approximation was considered sufficient. During the seventeenth century, the effort of coordinating the proliferating plant names found in different botanists' works got under way; Caspar Bauhin's *Pinax* (1623) was a milestone in compiling the various synonyms then in circulation. Comprehensive systems of classification had to wait a further generation; the most celebrated was Tournefort's, expounded in his *Institutiones Rei Herbariae* (1700). Tournefort's artist, Claude Aubriet, pioneered the use of dissections to convey additional information about fine structure.

Beginning in the 1730s, Carl von Linné (Latinized as Linnaeus) proposed a system of classification based on the numbering of the plant's sexual organs; the families were named Monandria, Diandria, Triandria, etc., depending on the number of stamens. During the heyday of this system, in the latter half of the eighteenth century, many artists were satisfied by portraying the flower alone, relegating the rest of the plant's anatomy to a subordinate position (literally, as in the case of Sowerby's *Helianthus* in plate 24, where the leaf is rendered in outline behind the coloured detail of the flower). While such plant portraits ideally included a dissection of the flower, many eminent

Ink drawing by Claude Aubriet (1665–1742) of a dandelion, commonly called *Taraxacum officinale*, though the taxonomy is confused

works skimped on this requirement, and much that passed as botanical art in the later eighteenth century – including some of the more famous works of Redouté – might today be reclassified as flower painting.

During the late eighteenth century, Linnaeus's system of classification came under attack from botanists who wanted to see a 'natural classification' in place of his artificial one; Adanson and Jussieu in France, Candolle in Switzerland, and Lindley in England, were among the major figures of this movement. In line with the demands of these botanists, their artists began to depict the entire plant, on the principle that it could not always be known in advance what might turn out to be the diagnostic characters. The interest in the morphology of plants that surged to prominence in this movement is shown in one of the period's masterpieces, the *Leçons de Flore* of Turpin and Poiret (see plate 35).

By the second quarter of the nineteenth century, a fairly consistent standard of botanical illustration was being offered by such publications as *Curtis's Botanical Magazine* and the *Botanical Register*, in which flower structure, leaf and stem morphology, and details of other relevant structures were shown. Such depiction, nonetheless, had its critics. In the last quarter of the century, the gardening journalist William Robinson made great claims for a new departure in plant portraiture, associated with his favourite artist, Henry George Moon (see plate 46). Moon's drawings did not include dissections, and often failed to convey the sort of information that a botanist required for precise identification, but Robinson praised his work for its accurate rendition of the habit of plants as they appeared in the garden, something which could not always be inferred from illustrations based on herbarium specimens. This elevation of horticultural over botanical standards in art had a great influence in England, and much early twentieth-century plant illustration in this country showed Moon's influence.

There is a separate tradition of horticultural painting for use in the identification of cultivated varieties. Where different varieties of one species need to be distinguished, it is permissible for the artist to ignore those factors that the varieties share, in order to concentrate on those differences which define the variety. The Hooker fruit drawings are one example: there has been no attempt to depict the entire tree, only those details – fruit, bud, etc. – that distinguish, say, 'Bell's Scarlet' apple from 'King of the Pippins'. Similarly, in the composite plate of orchid drawings, it is usually only the flower that is represented, since it is the improvement of the flower that is the principal aim of orchid hybridists.

Botanical art in the field and the studio

MOST BOTANICAL ART is made in the studio, rather than in the field. What can be made in the field is rough sketches, from which a finished portrait can be made later. Two collections of such drawings in the Society's library are cases in point. Frederick W. Burbidge kept a scrap book of drawings while travelling in Borneo and the South Seas for the Veitch nurseries; his drawings are rough sketches to assist in the later identification of specimens brought back to England. Reginald Farrer's drawings are frankly clumsier, despite a higher artistic aspiration shown in his incorporation into many of his drawings of glimpses of the scenery in which he found his plants.

In some cases, the artist may have only a herbarium specimen to work from; Walter Hood Fitch exulted in the challenge posed by such work: 'Sketching living plants is merely a species of copying, but dried specimens test the artist's ability to the uttermost; and by drawings made from them would I be judged as a correct draughtsman'.

The portrayal of garden cultivars in the present century has tended, like botanical art in the eighteenth, to be largely carried out in the private sector. Two major sources have been nurseries, and private growers who have commissioned portraits of award-winning plants. Before Iris Humphreys became the Society's orchid artist, for example, she had already painted orchid hybrids for the nurseries of Charlesworth, and Armstrong and Brown; and most of the drawings of J.L.Macfarlane in the Society's collection were originally made for the exhibitors of prizewinning orchids.

Botanical nomenclature

IN MANY OF THE PLATES in this volume, the name given to the plant in the text differs from that which the artist originally wrote on the drawing. This is particularly the case with the long polynomials that characterized plant names before the mid-eighteenth century, an example of which can be seen in Ehret's sweet-pea portrait (plate 9): *Lathyrus distoplatyphyllus hirsutus mollis, magno et peramoeno flore odoro*. Let me take this as my example in what follows.

Linnaeus proposed a reform of botanical nomenclature, in which, for purposes of handy reference, the long descriptive name would be replaced by a two-word code, consisting of the generic name – in our example, *Lathyrus* – and an epithet which would indicate which species within that genus was being discussed – in our example, *odoratus*. Instead of nine words, we now have a simple binomial: *Lathyrus odoratus*. This name is understood to be, not a description of the plant, but simply an identification tag; it does not even matter if the epithet is inaccurate (how many plants with the epithet *indicum* are actually native to India?); if the reader wants to identify the plant, he should consult a full description.

In drawing up his classification of plants, Linnaeus played fast and loose with the existing nomenclature. He was happy to reassign traditional Latin names to new genera, so that *Lonicera*, formerly a name for mistletoe, became the generic name for honeysuckles, and *Ilex*, the classical name for the holm oak, became the generic name for hollies. Frequently, the older Latin names remained as specific epithets: so in the cases of the common ivy, *Ilex aquifolium*, the holm oak, *Quercus ilex*, and the bog myrtle, *Myrica gale*, the epithet records the traditional name.

Despite objections at the time, Linnaeus's nomenclature (unlike his system of classification) has become accepted internationally as a standard, with 1753, the date of the publication of his *Species Plantarum*, becoming the official date for the beginning of modern plant names. No earlier name is valid, unless Linnaeus adopted it; one consequence is that names which were introduced by his predecessor Tournefort, but rejected by Linnaeus, are now officially credited to later botanists, like Adanson and L'Héritier de Brutelle, who revived them during the breakup of the Linnaean system of classification. Anyone who wants to examine pre-Linnaean plant illustrations must be prepared for hours of happy (or not so happy) work trying to deduce the modern names. The essential tool for this job is Hermann Richter's *Codex Botanicus Linnaeanus*, published in 1840: essentially an edition of Linnaeus's *Species Plantarum*, with all the previous synonyms that Richter could trace listed under each of Linnaeus' names, and a good cross-index.

The acceptance of Linnaeus as a standard has not, however, stopped plant names from changing in the two and a half centuries since. There are several reasons for subsequent name changes. First and most important of all, botanists are only human, and the pages of the *Index Kewensis* – the directory listing the first publications of all botanical names for the higher plants – are like a battlefield strewn with the corpses of superseded notions of classification. Botanists tend to be either lumpers or splitters, given either to the creation of inclusive categories or to the fine discrimination of variants, and either tendency can provoke a later revision. Many inclusive genera have been subdivided by later botanists (as has recently been the case with the horticulturally important genus *Chrysanthemum*, which has been broken up into fifteen different genera); other genera have been amalgamated (as with *Tacsonia* – see plate 44 – which has been absorbed into *Passiflora*).

A second major reason for name changes is the well-known phenomenon of multiple discovery and description. No botanist has read all the previous literature, and over and over again botanists have described plants in ignorance of the fact that they have been described before. The most effective way to cope with this problem, and the acrimony caused by the collision between two names, each of which has been in use for a period in different countries (think of *Sequoia* and *Wellingtonia*, *Dahlia* and *Georgina*, *Poinsettia* and *Euphorbia pulcherrima*), has been to adopt the first published description as the one that determines the name. This rule, informally recognized by botanists for decades, was formally adopted by the International Botanical Congress of 1866 (known as the Paris rules), and has since been tightened and refined, as successive Botanical Congresses have produced revised versions of the *International Code of Botanical Nomenclature*. One unintended consequence of the adoption of this rule has been that, as the earlier botanical literature is raked over, previously unfamiliar descriptions are unearthed, which under the rules must then be adopted as valid names, even if they overturn names accepted for a couple of centuries. And now that historically minded botanists are increasingly turning their attention to nurserymen's catalogues as sources of first published descriptions, there seems no end in sight to the process of change.

Botanical nomenclature is like common law: everything depends on precedent, and you can never be certain that a previously overlooked precedent will not be discovered that will change the situation. When the *Index Kewensis* was first compiled, in the 1880s, Sir Joseph Hooker and his colleagues confidently decreed which names were valid and which were synonyms; but since that time so many of these judgments have been overturned that the *Index* now largely confines itself to listing the published names, and letting the botanists argue over their status. It can take even an expert botanist much research to establish what ought to be the valid name of a plant.

In 1855, Georg August Pritzel published a work entitled *Index Iconum Botanicarum*, with a supplement added in 1866. In this he attempted to list the major illustrations of plants. In the 1920s, the Royal Horticultural Society decided to produce an updated version of Pritzel under the editorship of Otto Stapf, the former Keeper of the Kew Herbarium; it was published in 1930–31 under the title *Index Londinensis*, in six volumes, with two volumes of supplement in 1941. This work, following the *Index Kewensis*, took 1753 as its starting date, although it did include references to a few earlier works, such as Rheede tot Drakestein's *Hortus Malabaricus* (1678–1703) and Rumphius' *Herbarium Amboinense* (1741–50), whose illustrations it was known Linnaeus had used as types for plants in cases where he had no access to specimens. The lessons of the *Index Kewensis* had been learned by the 1920s, and Stapf made little attempt to adjudicate between rival classifications, often recording illustrations of the same plant under different names. This work is an essential tool for anyone working on botanical illustrations.

Nomenclature of cultivated plants

NO ATTEMPT WAS MADE in the *Index Londinensis* to index illustrations of garden varieties, although a certain number appear in its pages simply because of the nineteenth-century habit of giving garden forms Latinate names. Generally speaking, there is no index that lists cultivated varieties.

The naming of garden varieties was traditionally bedevilled by an uncertainty over their status. Variation in garden forms had been evident since ancient times, but early botanists were uncertain how to account for it, or even, in the absence of a species concept, to know what significance to ascribe to variation.

When new varieties arose in cultivation, whether by viral infection, by sporting, or as the result of accidental crosses, they were distinguished by descriptive names such as 'Blue bear's ear', 'Hair coloured bear's ear', and 'Greatest fair yellow bear's ear with eyes' (all names found in Parkinson's *Paradisus Terrestris*, 1629). When the number of varieties distinguished by fine differences became too great, fancy or arbitrary names began to be used instead, as in the case of tulips, which in the seventeenth century began to be given names like 'Semper Augustus'. Botanists, however, preferred to give varieties descriptive names until the Linnaean reform of nomenclature; and even Linnaeus allowed the use of trinomial names to distinguish subspecies and some varieties. Nurserymen in the nineteenth century had a tendency to greet any sustainable variety with a Latin name modelled on Linnaeus' trinomials, even where such varietal names indicated merely a juvenile form which lost its distinctive colouring on maturity.

Variation within a recognized category of plants was one thing; cross-breeding was another. Many botanists denied the possibility of hybridization until the eighteenth century, when Thomas Fairchild produced a cross between a carnation and a Sweet William. Extensive programmes of deliberate hybridization began about the end of the eighteenth century, with Rollisson's nursery in south London experimenting with Cape heaths, Dean Herbert in Yorkshire with various bulbous plants, and Thomas Andrew Knight in Herefordshire with fruits. The greatest sensation came in 1856, when John Dominy, a foreman at Veitch's nurseries in Chelsea, produced the first hybrid orchid; it was only in the wake of his success that John Lindley and other botanists began to suspect that some plants they had regarded as species may have been natural hybrids.

At the International Botanical Congress of 1866, Augustin-Pyramus de Candolle recommended that henceforth all cultivated varieties be given names in a vernacular language in order to distinguish them from naturally existing species and subspecies, which received Latin names. It took a long time before this recommendation was generally adopted, and the accompanying orthography was not finalized until the International Botanical Congress of 1950, when William T. Stearn, representing the Royal Horticultural Society, drafted a Code for the Nomenclature of Cultivated Plants, which was subsequently adopted. The names of cultivated varieties, fancy names in a vernacular language, are now given in Roman type (to distinguish them from the Latin names which are italicized), and either prefaced by the abbreviation cv. or set within single inverted commas.

This Code introduced the term 'cultivar' as an abbreviation for 'cultivated variety' – a variety which either originated in, or was only maintained in, cultivation. (So, for example, a form which is discovered in the wild but is kept distinct by being propagated for garden use would count as a cultivar.) A further term, 'grex', was introduced for series of cultivars of common ancestry – for example, the Russell lupins, or the Shirley poppies introduced by the Royal Horticultural Society's former Secretary, William Wilks, who let his parent collection pollinate each other without recording specific parentages. There is no poppy called 'Shirley', but a number of individually named cultivars grouped together under the collective name of Shirley poppies.

The complex nomenclature of orchid hybrids displays every problem that can be encountered, as shown in plate 70. First of all, extensive hybridization has taken place not only within, but between, genera. The early hybrid genera were given composite names indicating their ancestry, so that the result of crossing *Brassia*, *Laelia*, and *Cattleya* became known as *Brassolaeliocattleya*; but the complexity of hybrid genera has increased (as many as six genera may now be involved in a single hybrid), and it is now more customary to give a simpler arbitrary name to such artificial genera, naming them after some famous botanist or orchid grower, so that Gurney Wilson, for example, is commemorated in *Wilsonara*. Secondly, within each genus, there are now numerous grexes, each with its own range of cultivars. Accordingly, one of the orchids depicted in plate 70 is *Cymbidium* Magna Charta 'Spring Promise', with Magna Charta the grex name, and 'Spring Promise' the cultivar name.

Plant registration

EVER SINCE ROBERT THOMPSON'S work on fruit varieties in the early nineteenth century, the Society has tried to sort out the nomenclature of cultivated plants.

The Society's Orchid Committee was founded in 1889. In 1899 it began compiling a list of award-winning orchids, and later financed the work of R. A. Rolfe and C. C. Hurst in producing the *Orchid Stud Book* (1909), the first attempt to list the parentages of any category of garden hybrids. By then, the famous orchid nursery of Sander of St Albans had begun compiling hybrid lists, producing a cumulative edition in 1945. In 1949, finding the production of the list a financial burden in their straitened postwar circumstances, Sanders approached the Society for assistance, and thereafter the Society funded the project until in 1954 it took over the compilation of the list. Meanwhile, the Society's Narcissus and Tulip Committee had issued its first checklist of daffodil names in 1907.

One of the recommendations of the 1950 International Botanical Congress was the establishment of International Registration Authorities for different categories of cultivated plants. The Royal Horticultural Society is now the Registration Authority for eight categories of plants, more than any other single organization: *Dahlia*, *Delphinium*, *Dianthus*, *Lilium*, *Narcissus*, *Rhododendron*, orchids, and conifers.

Materials and techniques: paper

MOST OF THE DRAWINGS in this book were made on paper, with a certain number on card or board, which have the advantage of being less likely to crease on frequent handling. The type of paper generally used by European artists has changed over the centuries, however. Before the nineteenth century, almost all paper was made from linen and cotton rag. Coarser papers have more lumps and visible particles embedded in their surfaces; finer papers were traditionally hand-rubbed with a polished stone (later, machine-rolled) to yield a smooth surface. In order to make the paper less absorbent, so that ink and colours would not bleed into it, it was then usually dipped in some mixture of gelatine and alum. Alum, however, can produce sulphuric acid, leading to a brownish discolouration of the paper. Since the amount of alum used in the process varied according to local conditions, such as temperature in the paper mill, this could vary unpredictably, and successive sheets in the same album can discolour at different rates.

By the early nineteenth century, specially produced drawing paper was being manufactured using a wove process, which yielded a finer and more uniform surface. This technique was pioneered in the eighteenth century by the firm of Whatman, and many of the drawings in the Society's collection are on Whatman paper. By the middle of the nineteenth century, the demand for paper was exceeding the supply of rag, and rival processes were developed using other raw materials: first esparto grass, and later ground wood pulp, were the most successful. Most commercial paper manufactured today is made from wood pulp; one result is a sturdy drawing paper with a more uniform and finer white surface than was ever previously available. The drawback is that the process of making the pulp leaves the paper with a high acid content, that increases the risk of discolouration and deterioration.

Other sorts of paper can be found in the collection. Most of the Chinese and Japanese drawings in the collection were made on a very thin rice-based paper, which is difficult to handle safely without tearing. Many of F. W. Burbidge's scrapbook drawings were made on tracing paper, a paper made transparent by brushing it with oil of turpentine or a similar substance; the long-term difficulty with tracing paper is its increasing brittleness with age, and some of Burbidge's drawings eventually sheared along the creases where they were folded.

Some drawings in the collection have been made on vellum (calf or goatskin, specially stretched and shaved), an exceptionally hard-wearing material for the purpose. The special copy of Turpin and Poiret's *Leçons de Flore* made for Louis XVIII was printed on vellum, and its use as a drawing surface continues to the present day.

Materials and techniques: pigments

MOST BOTANICAL ARTISTS have tended to work in watercolour, for a variety of reasons. First, the artist has needed to work quickly. Although it might seem that plants make good stationary models, it is in fact difficult to train a plant to sit patiently for its portrait: flowers wilt and fade, they change colour, they can change their orientation during the course of the day, and sometimes the artist has only a few hours before the appearance of the specimen alters irreversibly. (This was especially the case before the days of the refrigerator, which has made it possible to prolong the sitting period of some specimens.) The speed with which watercolours dry, the facility with which they can be mixed, and the ease with which they can be carried, have long recommended their use for plant portraiture.

Further, since oil paint is built up in layers, the surface of an oil can become brittle in a way that watercolour seldom does. For this reason oil paint is seldom used on paper; less flexible materials like board and canvas are preferred. Some of the earliest artists in this volume, Pieter van Kouwenhoorn and Claude Aubriet, used oil on paper or vellum, and some of their pictures have suffered damage from the chipping of pigment. For all these reasons, botanical art in oils has been a comparative rarity. Raymond Booth's work (see plate 65) attracted much interest when he began exhibiting in the 1950s, precisely because his use of oils was such a novelty in botanical art.

The porcelain painters who drew W. F. M. Copeland's daffodils used gouache, a watercolour made opaque by mixing with white (plate 47), and the Chinese and Japanese artists whose work is represented in the collection used what was basically watercolour, sometimes stiffened with gum. Other artists have used ink or pencil for monochrome work. One group of drawings, by F. W. Burbidge, adds further conservation problems to the curator's burden: as part of an endeavour to depict daffodil bulbs, he glued pieces of bulb tunic onto his drawings.

The artist has at his disposal only such colours as his pigments can make available; in other words, the colouring of a drawing is always an approximation, or a selection, of the colours of the real plant. Most traditional pigments were derived from plant and animal substances (e.g. cochineal) or from certain minerals (e.g. cobalt yellow), and the artist bought them in ground or powdered form and mixed them together, or even ground them himself. Since the mid nineteenth century, the range of commercially prepared pigments has greatly increased, and even where the names are the same, the pigments available today are not physically identical with those that an eighteenth-century artist would have used.

And have those pigments lasted to the present day without changing? Watercolour in particular fades on much exposure to light, and several pigments are fugitive. Some drawings in the Lindley Library, especially the Reeves Chinese drawings, show the deterioration of pigment, most commonly

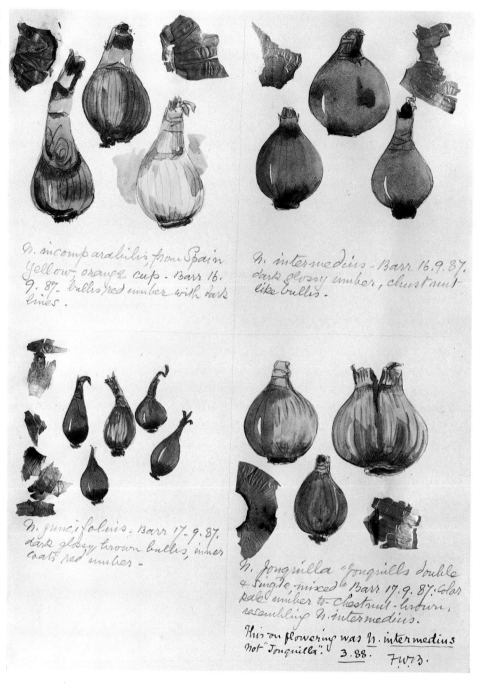

where the old lead-based white pigment has turned a dark grey on interaction
with the smokier atmosphere of an earlier day.

Quite apart from the question of choice of pigments, the ways in which
artists have attempted to represent the colours of plants have varied over the
centuries. While flower painters such as Van Huysum delighted in their
attempts to represent the translucence and light reflections of a dewdrop on

26

a petal, botanical artists of the eighteenth century tended to restrict themselves to a depiction of the body colour of a given area, dismissing the effects of light as an irrelevance to the question of identification. In the early nineteenth century, on the other hand, an increased interest in the nature of plant tissues went hand-in-hand with an artistic interest in capturing the different textures of plant surfaces: the use of white highlights to indicate points of high reflection, whether on the curve of a petal or on the surface of an apple; Barbara Cotton's unrivalled rendering of the down on a peach skin; possibly even, in the case of Augusta Withers, the use of an enamel undercoat to suggest the translucence of a gooseberry skin. An increased use of chiaroscuro also characterized this period: artists such as Withers and Turpin deliberately heightened contrasts of colour and shade in their drawings, in order to emphasize detail. By the beginning of the present century, on the other hand, English artists were preferring to use paler colours and reduce the degree of chiaroscuro in their work.

The printing of illustrations

IT SHOULD NEVER be forgotten that the primary purpose of a botanical illustration is to be printed. No matter how accurate the representation of a plant, its use to botanists is restricted unless it is widely distributed, so that botanists anywhere can use it. An individual drawing can only be used by those who visit the particular collection in which it is held; even a printed book, if published in a small number of copies, can be of limited use. A case in point was the great *Flora Graeca* of John Sibthorp, for which Ferdinand Bauer made the illustrations; only thirty copies were originally published, and a botanist like Tenore had to travel from Naples to Paris when he needed to see a copy.

Of the illustrations in this volume, fourteen (plates 21, 23, 25–30, 36, 44, 46, 50, 67, 69) are original drawings for engravings or lithographs that were subsequently published. Various others were probably intended for publication, although in the event not so used.

The earliest published botanical drawings were printed from woodblocks carved so that the outlines of the figure projected from the block. This method placed a certain upper limit on the degree of detail an illustration could offer, although Mediterranean woodblock artists by the late sixteenth century could create some very complex images. From the beginning of the seventeenth century, woodcuts were gradually superseded by engravings, in which lines were incised into a metal sheet; this technique greatly increased the amount of detail that could be depicted. In England, line engraving remained the dominant technique, and English drawings accordingly are strong in line, with emphatic outlines and shaded areas indicated by multiplied lines or cross-hatching. In France, on the other hand, stipple engraving, using a multitude of small dots (a technique ironically pioneered in England), had become the norm by the early nineteenth century; so an artist like Bessa (plate 36) played down the importance of line in favour of more finely modulated colour, relying on the engraver to render the effect by increased stippling.

Unfortunately, printed copies do not always convey accurately all the information that the original drawing holds, and the colouring of the plant is the information least reliably conveyed. If the colours of a drawing are an approximation of the colours of a real plant, the colours of a printed illustration are a further approximation of the colours of an original drawing, especially since the introduction of mechanical methods of colour printing.

Printed illustrations were seldom coloured by the original artist; colouring was usually applied later, under the direction of the owner of the volume. Sometimes a certain number of copies of a work would be coloured by the artist, or at least under the supervision of the publishing house, in order to ensure a consistency of colouring. Nonetheless, until the end of the eighteenth century, it is unwise to rely on the accuracy of colouring of a printed illustration, unless its circumstances of colouring are documented.

Artists and publishers tried various ways of getting around this difficulty. In Ehret's later work (see plate 8), there is a tendency to simplify the colouring into broad masses of one tint or shade, so that colourists copying the drawing would have as simple a job as possible. At the end of the eighteenth century, Rudolph Ackermann and some rival publishers found a way of ensuring uniformity by using an assembly line of colourists, each of whom would be given a particular colour to put in a specified place on each copy; as a result, with publications like *Curtis's Botanical Magazine*, inconsistency of colouring became almost a thing of the past. It was always watercolour which was used for colouring engravings, since the use of any opaque colour would efface the engraving lines.

The advent of lithography, a technique invented by Alois Senefelder in 1798, at first changed little. The word 'lithography' means 'stone writing', because the original drawing surface was a polished block of stone, though later zinc and aluminium panels were increasingly favoured. An image was drawn on this surface with a greasy medium – usually chalk or crayon; the stone was then moistened, and printer's ink applied with a roller; the ink adhered to the greasy medium, but not to the water. Paper was then placed on the stone, and a scraper run over it, transferring the ink to the paper. Artists like Ferdinand Bauer were delighted with lithography, and it quickly came into use for botanical illustration. One of the advantages of lithography was that it enabled the artist to draw directly on the printing surface, without the need of a special craftsman to transfer his drawing; Walter Hood Fitch, in particular, often drew directly on the stone. Lithographs were originally monochrome, and had to be coloured in the same way as engravings; *Curtis's*

Botanical Magazine switched from engraving to lithography in 1845, but went on being coloured by hand until 1949.

By the 1840s many lithographers were experimenting with ways of printing in full colour, and by mid-century more than one version of 'chromolithography' was available. By the turn of the century, however, this in its turn was being ousted by screen printing: light, shone through a negative, was projected through a half-tone screen to produce a pattern of fine dots on a sensitized plate, which could then be used for printing. The early colour plates produced by this process were printed in three colours; today, a four-colour process, as used in this volume, is the norm.

Today, almost all commercial printing, of text as well as of illustrations, is done from photographic film. It is common for book illustrations to be assembled for photography by cutting out details and arranging them on a sheet; but this is in fact a very old practice, and there are drawings of Claude Aubriet's from the early eighteenth century on which bits of paper with extra details have been pasted on.

Photography

MANY PEOPLE, on first being introduced to the exigencies of botanical art, ask whether modern technology has not rendered the artist obsolete: has photography not superseded drawing as a means of recording plants?

Photography, even in its present improved forms, presents certain problems in making a plant portrait. First of all, there is the question of focus: a plant is a three-dimensional structure, and the more closely one part is brought into focus, the less in focus will the rest of the plant appear. Secondly, problems of accuracy in colour bedevil photography just as much as working with pigments; given a plant with dark bark and light flowers, the more the lighting is adjusted to render the detail of the one, the more the detail of the other will be obscured. This problem is less apparent under studio lighting than in the wild, and, with the publication in 1977 of *Wild Flowers of Britain*, Roger Phillips pioneered a new form of plant photography, with specimens, evenly lit, laid out as on herbarium sheets. Nonetheless, a large number of photographs is necessary to convey the information that can be provided by a single well-presented drawing, and the artist has the advantage of being able to manipulate the image in order to emphasize the features that are considered to be of diagnostic importance.

The photographer, then, has yet to supersede the botanical artist, and for the foreseeable future the sort of illustration represented in this volume will continue to play an important role in both botany and horticulture.

Bibliography

Anon. 'Meet the RHS orchid artist: Cherry-Anne Lavrih, Dip. AD, ATC', *Orchid Review*, 98 (1990), 277–81.

Blunt, Wilfrid. *The Art of Botanical Illustration*, revised and enlarged by William T. Stearn, Woodbridge, Suffolk, The Antique Collectors' Club, 1994.

Bridson, Gavin D. R., & Wendel, Donald E. *Printmaking in the Service of Botany*, Pittsburgh, PA, Hunt Institute for Botanical Documentation, 1986.

Desmond, Ray. *A Celebration of Flowers: Two Hundred Years of Curtis's Botanical Magazine*, Royal Botanic Gardens, Kew / Collingridge, 1987.

Elliott, Brent. 'The printing of botanical illustrations', *The Garden (Journal of the Royal Horticultural Society)*, 118 (1993), 155–7, 220–22.

Feller, Robert L. (ed.) *Artists' Pigments: a Handbook of their History and Characteristics*, Cambridge University Press, 1986 to date.

Fletcher, H. R. *The Story of the Royal Horticultural Society 1804–1968*, Oxford University Press, 1969.

Hills, Richard L. *Papermaking in Britain 1488–1988: a Short History*, London, Athlone Press, 1988.

Krill, John. *English Artists Paper: Renaissance to Regency*, London, Trefoil Publications, 1987.

Rittershausen, Wilma. 'The orchid artists of the Royal Horticultural Society', *Orchid Review*, 95 (1987), 148–51.

Roach, F. A., & Stearn, W. T. *Hooker's Finest Fruits*, London, Herbert Press, 1989.

Stearn, William T. *Flower Artists of Kew*, London, Herbert Press, 1990.

Tjaden, W. 'The loss of a library', *The Garden (Journal of the Royal Horticultural Society)*, 112 (1987), 386–8.

Tjaden, W. 'The Lindley Library of the Royal Horticultural Society, 1866–1926', *Archives of Natural History*, 20 (1993), 93–128.

The Plates

Note: In giving the dimensions of the drawings, all measurements
have been rounded to the nearest half-centimetre.

PLATE I

Pieter van Kouwenhoorn (*fl.* 1620s–1630s)
Paeonia officinalis   Paeoniaceae

This drawing is reproduced from an album of forty-six coloured drawings on paper, bound in vellum, with a manuscript title page inscribed 'Verzameling van bloemen naar de natuur geteekend door [Collection of flowers drawn from nature by] Pieter van Kouwenhoorn'. Kouwenhoorn (also written Couwenhoorn) was a glass painter, working in the 1620s and 1630s in Haarlem and Leiden; he was best known for his windows in the Annahofje in Leiden.

This work was obviously prepared with publication in mind, for the plant names are often given in Latin, French, and German. On some pages lines have been ruled for names that have not been written in, as in the case of the passion flower, the most recent introduction at the time of drawing. Altogether, some 200 plants are depicted, as well as slugs and insects.

Peonies (the name is sometimes encountered in the nineteenth-century fake Latin form 'paeony') were first introduced into England from southern Europe in the sixteenth century. By Parkinson's time, *Paeonia officinalis* was already being grown in English gardens in white and red forms, both single and double. It was not until the introduction of Chinese species in the nineteenth century that an extensive range of garden hybrids was developed.

Oil on paper, 45.5 × 32 cm
PROVENANCE: Reginald Cory bequest, 1936

PÆONIA

Piuoine

Pfingſtroſe

PLATE 2

Pieter van Kouwenhoorn (*fl.* 1620s–1630s)
Tulipa cultivars   Liliaceae

The first known tulip in Europe flowered in 1559 in the garden of Heinrich Herwart in Augsburg; Herwart had probably received it from Augier Ghislain de Busbecq, the Austrian ambassador at Constantinople, who had seen tulips growing in Turkey, where they were eagerly cultivated and fetched high prices. Before long, they were fetching high prices in Europe also. They were being grown in England by the time William Turner published his *Herbal*; by 1629, John Parkinson could list 140 varieties in his *Paradisus Terrestris*. Most tulip breeding, however, took place in Holland, where by the 1630s a single bulb of a specially prized flower, 'Semper Augustus', could sell for nearly 5000 florins, plus horses and carriage. In 1637, the tulip market suddenly crashed, and prices never again rose to such extreme levels. The popularity of tulips, however, both as garden flowers and as florists' or exhibition flowers, continued unaffected, and in England during the 1830s and 1840s there was a further surge of high prices, with 'Victoria Regina' being sold for £100 a bulb in 1837.

The oldest tulips have long since disappeared, although it is still possible to find some seventeenth-century cultivars. This tulip plate by Kouwenhoorn was probably drawn during the heyday of tulipomania; unfortunately, none of the varieties is named.

Oil on paper, 45.3 × 32 cm
PROVENANCE: Reginald Cory bequest, 1936

PLATE 3

Pieter Holsteyn (c.1614–1673)
Acanthus mollis   Acanthaceae

In the Middle Ages, lacking any notion of the geographical distribution of plants, men tried to identify in their native flora the plants they knew from their textbooks (descended, through generations of copying, from the writings of Theophrastus and Dioscorides, who had written about the plants of the Mediterranean world). The acanthus is native to southern Europe, but since *Acanthus spinosus* has the reputation of being the source of the ornament on a Corinthian capital, the concept of 'acanthus' was familiar in parts of Europe where the plant did not grow. So when 'acanthus' appears in a plant list by Alexander Neckam, about 1200, it is almost certainly some other plant that Neckam thought matched the description: John Harvey has suggested that it was actually the hogweed, *Heracleum sphondylium*. *Acanthus mollis*, the species shown here, was not introduced into England until about 1548.

This drawing is selected from an album of 142 coloured drawings bearing the manuscript title 'Flores a Petro Holsteyn ad vivum depicti'. Pieter van Holsteyn the younger was born in Haarlem around 1614, and died there in 1673. He became a member of St Luke's Guild in 1662. One drawing in this album, depicting a tulip, bears the inscription 'A.C. fe 1641'; this must be the work of Anthony Claesz (c.1607–49), on whose work Holsteyn probably modelled his style.

Watercolour on paper, 36.5 × 21 cm
PROVENANCE: Reginald Cory bequest, 1936.

Pl. fe.

PLATE 4

Claude Aubriet (1665–1742)
Punica granatum   Punicaceae

The pomegranate is native to Iran and Afghanistan, but was introduced to Europe during the Middle Ages. Albertus Magnus recommended its planting in monastic gardens in the middle of the thirteenth century; a century later, Henry Daniel's manuscript records that he could grow it in England, but not obtain fruit. By the time of Gerard and Parkinson, it was more widely grown in England, and specimens have occasionally fruited. It has more generally been treated as an ornamental plant, and in the eighteenth century there were six forms in cultivation, distinguished by the colour and pattern of their flowers: Philip Miller praised the double red form as 'one of the most valuable flowering trees yet known'.

Claude Aubriet was born in Chalons-sur-Marne in 1665. While working in Paris, he attracted the attention of the great botanist Tournefort, who commissioned him to produce illustrations for his *Institutiones Rei Herbariae* (1700). From 1700 to 1702 Aubriet and Tournefort travelled together in the Near East. Aubriet went on to become the official artist of the Jardin des Plantes, a position he held until his death in 1742. This painting is taken from a volume which bears on its boards the arms of the French royal family. It includes one of the earliest European illustrations of the coffee plant and of the vanilla orchid.

Oil on vellum, 42 × 30 cm
PROVENANCE: Reginald Cory bequest, 1936

PLATE 5

Anonymous
Tulipa clusiana × *T. schrenkii 'Semiplena'*　　　　　　　　Liliaceae

This tulip portrait is taken from an album of thirty-three coloured drawings of tulips, irises, and other bulbous plants, by an anonymous artist. The name Gottfried Läubli appears on an endpaper, but his identity has not yet been established. Some of the drawings have been annotated in an eighteenth-century hand, giving Latin names from Dillenius' *Hortus Elthamensis*; the names of some of the tulips are written in faint pencil at the top of some pages. In the same hand, the name Paolo Retti appears on an endpaper, presumably as the owner rather than the artist of the drawings.

Retti (1691–1748) was a court architect in Swabia in the 1720s. After two years in prison for embezzlement because of the delays in completing one of his buildings, he enjoyed a period of favour under the patronage of the Duke's financial adviser, Josef Süss Oppenheimer; but after the Duke's death, Süss was tried and executed on trumped-up charges, and Retti fled to Italy, his property confiscated by the court. (Perhaps this album was one of the things he had to leave behind?) As a result of this, Retti has a walk-on part in Lion Feuchtwanger's novel *Jew Süss*, where he is glimpsed at the gaming-tables on the night of the Duke's death.

The name of this tulip has, like some of the others in this volume, been cropped by a previous binder. The cultivar has been identified by Sam Segal, author of *The Tulip Portrayed* (1992).

Watercolour on paper, 44 × 27.5 cm
PROVENANCE: Reginald Cory bequest, 1936

PLATE 6

Michiel van Huysum (*c.*1704–1760)
S.M. van der Polle (dates unknown)
Flower and fruit studies

These two illustrations are representative of the tradition of flower painting, as it developed during the eighteenth century. Often at this period there was little to distinguish flower painting from true botanical art except the absence of dissections.

Michiel van Huysum was the son of Justus van Huysum (1659–1716) by his second wife, and brother of the more famous Jan van Huysum. Most of Michiel's dated work is from the 1750s; he died in 1760. At first the van Huysum brothers worked in association, but as Jan's style changed – with greater visual depth and well-developed backgrounds – he detached himself from the rest of his family and kept his techniques secret. Michiel was never to rival his brother's reputation. This drawing, dated 1755, shows *Nigella damascena* between two roses.

Nothing is known about S.M. van der Polle, the artist of the fruit and nut study shown below, depicting a peach, a squash, and a walnut.

Watercolour on paper, respectively 19.5 × 29 and 37 × 25 cm
PROVENANCE: Reginald Cory bequest, 1936

PLATE 7

P. van Weinsel
Fritillaria imperialis   Liliaceae

Crown imperial is a translation of *corona imperialis*, a Latin name which has been explained as an allusion to the Habsburg court. Certainly it was in Vienna, to which it was introduced by Clusius in 1576, that it was first cultivated in Europe. John Parkinson chose it as the first flower to discuss in his *Paradisus Terrestris* (1629): 'The Crowne Imperiall for his stately beautifulness, deserueth the first place in this our Garden of delight.' At that time no garden varieties were known, distinct from the species; Parkinson dismissed rumours of white-flowered forms. In the eighteenth century, Philip Miller listed eight cultivars, which were still extant less than a century ago.

This drawing is signed by P. van Weinsel; nothing is known about the artist. In addition to the fritillary, it depicts a guelder-rose, plums, raspberries, and butterflies. The dewdrops on the fritillary's petals are an attempt to imitate Jan van Huysum, who was celebrated for his skill in rendering droplets, and notorious for refusing to reveal how he did it.

Watercolour on paper, 37 × 27 cm
PROVENANCE: Reginald Cory bequest, 1936

PLATE 8

Georg Dionysius Ehret (1708–1770)
Ribes americanum   Grossulariaceae

Ehret was born at Heidelberg in 1708, and was apprenticed in the garden of the Margrave of Baden at Karlsruhe. In 1736 he began drawing plant portraits for George Clifford of Hartecamp, which were used in Linnaeus's *Hortus Cliffortianus.* Shortly thereafter he moved to England, where he spent the rest of his career, working at the Chelsea Physic Garden, the Oxford Botanic Garden, and for patrons such as the 3rd Earl of Bute and the Duchess of Portland. In 1768 Mrs Delany recorded that he was busy painting native English plants for the Duchess, and complaining about the effect that the microscopic examination of plants was having on his eyesight; some of these drawings are now in the Royal Horticultural Society's possession. Ehret illustrated Patrick Browne's *Natural History of Jamaica,* and collaborated with J. C. Trew on *Plantae Selectae* and *Hortus Nitidissimis.* By the time of his death in 1770 he was widely regarded as the best botanical artist of his day.

Ribes americanum, the American black currant, was one of the many eastern American plants introduced into England by Peter Collinson during the first half of the eighteenth century. It was growing at the Chelsea Physic Garden by the time Philip Miller published the first edition of his *Gardeners Dictionary* in 1731. It is sometimes recommended for autumn colour.

Watercolour on vellum, 53 × 37 cm
PROVENANCE: Reginald Cory bequest, 1936

PLATE 9

Georg Dionysius Ehret (1708–1770)
Lathyrus odoratus Leguminosae

In 1697, Franciscus Cupani published his *Hortus Catholicus*, including a description of a flower he had found growing wild in Sicily. Two years later he sent seeds to the botanist Robert Uvedale, and by the 1720s it was commercially available in England. Thomas Fairchild referred to it as the 'sweet-scented Pea', a phrase which was reduced to 'sweet pea' later in the century. The original sweet pea was a purple and maroon bicolour.

Cupani named the plant *Lathyrus distoplatyphyllus hirsutus mollis, magno et peramoeno flore odoro*. Linnaeus, in his *Species Plantarum* of 1753, renamed the plant *Lathyrus odoratus*, distinguishing two subspecies – Cupani's form, native to Sicily, and another form from Ceylon. No Sri Lankan sweet pea has ever been discovered, and even the Sicilian origin has been questioned: early eighteenth-century herbarium specimens resemble the modern cultivated forms more closely than any Mediterranean species. The recent suggestion that the original sweet pea was of South American origin, imported to Sicily by the Spanish, has not been accepted.

This drawing is dated 1757. Even though Ehret had been associated with Linnaeus since the 1730s, it bears the polynomial name that Commelin had originally published.

Watercolour on vellum, 49.5 × 35.5 cm
PROVENANCE: Reginald Cory bequest, 1936

PLATE 10

Simon Taylor (1742–1772)
Lythrum salicaria Lythraceae

Simon Taylor was born in 1742. As a youth, he won prizes from the Royal Society of Arts, and in 1760 was employed by Lord Bute to paint rare plants at Kew. He worked closely in association with Ehret, and they collaborated on some drawings; it is often difficult to tell which of the two artists was responsible for an unsigned drawing. On Ehret's death, the botanist John Ellis named Taylor in a letter as a promising young artist. He was also employed by the doctor and pioneering horticulturist John Fothergill, whose Taylor drawings were later sold to the Empress of Russia for £3,000. His date of death has been given variously as 1772, 1790, and c.1796. The Society's Taylor drawings formed part of an album formerly belonging to the Duchess of Portland.

The purple loosestrife is a native of Britain and Europe, growing abundantly in marshy places. As the name indicates, it was recommended by the early herbalists as a remedy for diarrhoea and dysentery, but its use as a medicinal plant declined during the eighteenth century. Stephenson and Churchill wrote in their *Medical Botany* (1836): 'Though it has long been celebrated in Ireland, it is seldom prescribed in regular practice. It has been given generally in the form of a decoction, made by boiling one ounce of the dried herb in a pint of water, down to half a pint. Of this dose may be three or four ounces twice a day.'

The purple loosestrife was introduced to North America as an ornamental flower in the nineteenth century, and has become naturalized, moving steadily westward across the continent. It has now reached the Pacific Coast, and in 1993 a campaign was launched in British Columbia advising people to 'get to know this plant, report it to the Canadian Wildlife Federation, hack it down, rip it up'.

Watercolour on vellum, 25 × 16.5 cm
PROVENANCE: Reginald Cory bequest, 1936

PLATE 11

Anonymous, attributed to Thomas Robins (1715–1770)
A fungus group

Margaret Cavendish Bentinck, 2nd Duchess of Portland (1715–85), was an amateur botanist, a friend of Rousseau, and, during the late 1760s, a patron of both Ehret and Simon Taylor. In 1786, her collection of drawings was sold at auction. Reginald Cory eventually came into possession of an album containing some of these drawings, which he bequeathed to the Society. In addition to the Ehret and Taylor drawings, it contained three drawings of fungi and butterflies, one of which is shown here.

None of the three is signed, but on stylistic grounds they have been attributed to Thomas Robins the elder (1715–70), who has recently become famous for his rococo garden scenes. Robins is known to have worked for the Duchess, and if the fungus drawings are of the same date as the Ehret and Taylor drawings from the same album (late 1760s), they could be his work.

The biology of fungi was slow in being understood, and it was not until the second half of the eighteenth century that botanists began to devote entire books to the subject. The first illustrated fungus books in England were James Bolton's *History of Fungusses, Growing about Halifax* (1788–91) and James Sowerby's *Coloured Figures of English Fungi or Mushrooms* (1797–1803). It was not until the nineteenth century that the life cycle and reproduction of fungi were finally understood.

Watercolour on paper, 31 × 28 cm
PROVENANCE: Reginald Cory bequest, 1936

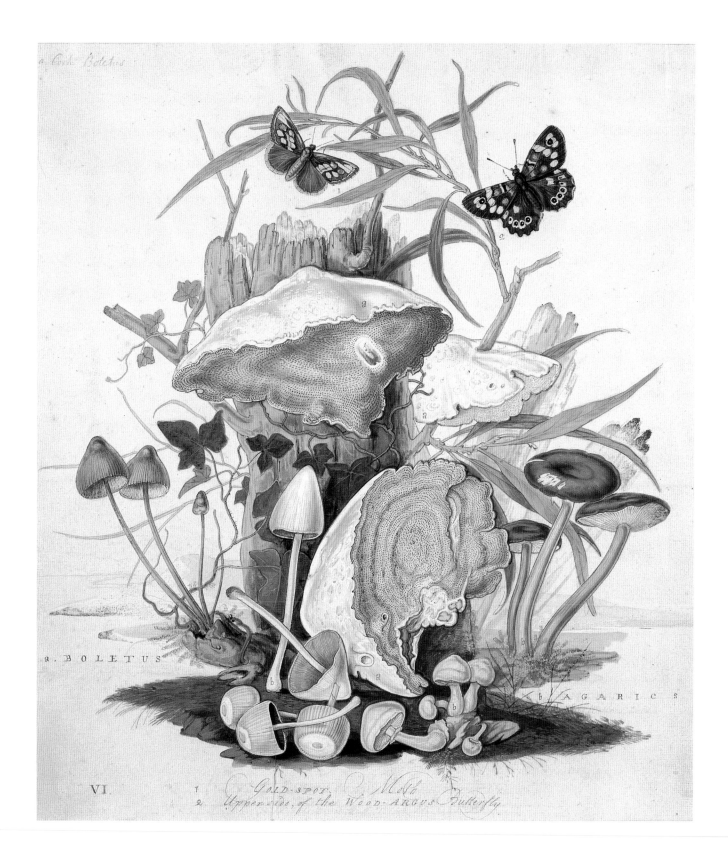

a. BOLETUS

b. AGARICS

VI.   1.   GOLD-SPOT   Moth

2.   Upperside of the WOOD-ARGUS Butterfly

ings of Asiatic plants by various (anonymous) hands. By the 1840s, these drawings were in the possession of the botanist Thomas S. Ralph (1813–91), who arranged them systematically in eight folio volumes with the title 'Flora Asiatica'. Ralph left England for Australia in 1851; the subsequent history of the drawings is unknown until they came into the possession of Reginald Cory, who bequeathed them to the Society.

The myrtle is native to western Asia: it was one of the first plants from this area to enter Britain, its introduction formerly attributed to Sir Walter Ralegh, and by the end of the sixteenth century six garden varieties were known to Gerard; double forms were around by Parkinson's time. It reached its peak of popularity in the second half of the eighteenth century, and was still one of the commonest of conservatory plants in the early nineteenth century.

Watercolour on paper, 44.5 × 28.5 cm
PROVENANCE: Reginald Cory bequest, 1936

PLATE 12

Anonymous
Myrtus communis   Myrtaceae

John Stuart, 3rd Earl of Bute (1713–92), was Prime Minister from 1761 to 1762, during which time he became known as the most hated man in England; his political career ended in disgrace, and much of his later career was devoted to botany and horticulture. He had become the administrator of the Royal Garden at Kew in 1757, resigning in 1772; during this time he employed artists such as Ehret and Taylor to draw plants. He built up a collection of over 2000 plant drawings, which were dispersed by auction in 1792.

Among these drawings was a collection of 737 draw-

PLATE 13

Anonymous
Pinus halepensis   Coniferae

In 1826, Prince Pückler-Muskau, on tour in England, noted with surprise the English enthusiasm for conifers; he himself was over-familiar with them from the Black Forest. This love of conifers, which is often thought to characterize the Victorians, had in fact begun in the seventeenth century, when the cedar of Lebanon was introduced. Trees with classical or Biblical associations, in particular, were greeted warmly, so that the pine shown here was first described (by Leonard Plukenet in 1696) as *Pinus Hierosolymitana*, the Jerusalem pine, and later generations came to call it the Aleppo pine.

Pinus halepensis (the name was coined by Philip Miller of the Chelsea Physic Garden) is a native of southern Europe and Asia Minor. It was introduced into England in the 1680s by the great nurseryman George London, who sent it to Bishop Compton at Fulham Palace. Insufficiently hardy to withstand severe winters, it is rare in cultivation in this country, and most of the early recorded specimens have died.

This unsigned drawing is taken from an album of drawings formerly in the Earl of Bute's collection.

Watercolour on card, 48 × 35 cm
PROVENANCE: Reginald Cory bequest, 1936

PLATE 14

Simon Felsted (*fl.* 1780s)
Bromelia pinguin   Bromeliaceae

The pinguin is native to Central America and the West Indies. Its date of introduction into Europe is not known exactly; it was being grown at Hampton Court in 1690, and the immigrant German botanist Johann Jakob Dillenius was growing it in his garden at Eltham before 1732. John Lindley, in his *Vegetable Kingdom*, said that it was used 'to destroy intestinal worms, and to promote the secretion of urine'.

Despite the deficiencies of the period's glasshouses, by comparison with the improvements that were to come in the nineteenth century, there was much interest in growing tropical exotics under glass in the seventeenth and eighteenth centuries. For example, a London gardener named John Cowell built a special house in Hoxton for his 'aloe' (*Agave americana*), and charged admission to see it; in 1730 he published a book about his collection of exotics, in part to attract the public back after he and the agave were injured in an attack by vandals.

Nothing is known about the life or career of Simon Felsted. This drawing is dated 1782, and came to the Society as part of Reginald Cory's bequest; Cory apparently bought it as part of a group with drawings, mostly of British orchids, by Thomas Robins and William Lewin, though whether this indicates a connection between the three artists in real life is not known.

Watercolour on paper, 36 × 25 cm
PROVENANCE: Reginald Cory bequest, 1936

PLATE 15

Thomas Robins the younger (1748–1806)
Himantoglossum hyrcinum   Orchidaceae

Himantoglossum (formerly *Satyrium*) *hyrcinum* – the name change was made by Sprengel in 1826 – the Lizard orchid portrayed here, has one of the most interesting histories of the native British orchids. According to Summerhayes (*Wild Orchids in Britain*, 1951), this species was first recorded in 1641, and was known only in Kent until about 1850, when it died out over its main distribution; over the next half-century, it was recorded sporadically in the southeast. Then, especially after 1919, probably because of the increasing warmth of the climate, it began increasing its range, until it could be found from Yorkshire to Devon. Since Summerhayes' time, however, it has declined again, and is now rare in many parts of its range; but this decline has been in part due to the publicity it attracts when spotted, and many specimens have been recorded in the act of digging them up. Perhaps global warming will increase its range once more.

Thomas Robins the younger was born and died in Bath, where his father had been a successful artist. The younger Robins spent much of his career teaching drawing. In 1787 he advertised 'pictures of exotic plants and insects' for sale. His last years were apparently spent in financial hardship; the year before his death, he wrote to the Bath and West of England Agricultural Society beseeching them to buy two volumes of his natural history drawings 'in aid of the artist's great Distress'.

Watercolour on paper, 39 × 25 cm
PROVENANCE: Reginald Cory bequest, 1936

Satyrium Hircinum.
Lin: **XX.....1.**

Lizard, *or* Great Goat Orchis.
I. Robins Del. Bath *1784.*

PLATE 16

Margaret Meen (?–1824)
Passiflora laurifolia   Passifloraceae

In 1571 Nicolas Monardes published an account of a curious South American plant which the Spaniards called 'Granadillo', in which could be seen 'figures which are thynges of the Passion of our lorde'. By the early seventeenth century, the plant was being called *Flos Passionis* (rendered into English as 'passion flower'), and the symbolism had been standardized:

leaves and tendrils = the hands and whips of the jailers
five stamens = five wounds
styles = the whipping-post
three stigmas = three nails
corona = crown of thorns

The date at which the passion flower was first grown in Europe is uncertain. According to Tobias Aldini, a specimen was planted in the Farnese garden in Rome in 1619; it was being grown in English gardens by the time of Parkinson's *Paradisus Terrestris* of 1629. Parkinson denounced the Passion symbolism, which could prove a point of contention as late as the nineteenth century, when the French language of flowers books gave the plant's meaning as 'Faith', while English (Protestant) language of flowers books gave its meaning as 'Superstition'.

This drawing depicts *Passiflora laurifolia*, the laurel-leaved passion flower or water lemon, and is dated 1785. It was first grown in England about 1690, by the Earl of Portland.

Margaret Meen exhibited at the Royal Academy from 1775 to 1805, and died in 1824. In 1790 she published two parts of a folio work, *Exotic Plants from the Royal Gardens at Kew*.

Watercolour on vellum, 30 × 22.5 cm
PROVENANCE: Reginald Cory bequest, 1936

H.M.1785. Passiflora. Laurifolia. Gynandria   Pentandri.
                                                                                tt.20.

PLATE 17

Elizabeth Smith [?]
Malvaviscus arboreus   Malvaceae

Among the drawings bequeathed to the Society by Reginald Cory were twenty-eight drawings on vellum, apparently by a group of sisters. Seven are signed Aug.S. [Augusta Smith], some of them forming part of a numbered sequence as though intended for book illustrations or at least a bound volume in a private collection. Nine are signed Em.S. [Emma or Emmeline Smith?], and twelve signed El.S. [Elizabeth or Eliza Smith?]. There are known artists of those two names at the end of the eighteenth century, but none who would have been active in the 1780s, when some of the drawings are dated.

Shown here is a drawing by El.S. The plant depicted is a native of Jamaica, first cultivated in this country by the Duchess of Beaufort at Badminton in 1714. Dillenius, in his *Hortus Elthamensis* (1732), called it *Malvaviscus arborescens*; Linnaeus absorbed it into his genus *Hibiscus*, making it *Hibiscus malvaviscus*; Adanson, in the 1760s, partially revived Dillenius' name, so that it is now *Malvaviscus arboreus*.

Watercolour on vellum, 35 × 26 cm
PROVENANCE: Reginald Cory bequest, 1936

PLATE 18

Pierre Ledoulx (1730–1807)
Hibiscus rosa-sinensis

Malvaceae

This drawing is one of ninety-one contained in two albums with the manuscript title 'Collection du Règne Végétal', mostly depicting plants grown in the garden of J. van Huerna, near Bruges, between 1792 and 1815. The drawings were collected and bound together by Huerna's grandson Joseph de Pelichy in 1831. The drawings are by four different hands: twenty-three are signed by Jean-Charles Verbrugge (1756–1831), fifteen by Pierre Ledoulx (1730–1807), two by Bernard Cromber (*fl. c.*1800), and two by Joseph François Ducq (1762–1829).

This drawing has been inscribed by Huerna, 'Hibiscus chinensis en fleurs à Averloo en 1796 dans mon Jardin'; although unsigned, it appears to be the work of Pierre Ledoulx.

As its name indicates, *Hibiscus rosa-sinensis* was thought to be a native of China, although it has never been found there in the wild; it is now thought to have originated in east Africa or in the Indian Ocean islands. In England, it has always required glasshouse cultivation; English growers have, as a result, been little involved in the production of new varieties. The first important hibiscus breeder was Charles Telfair of Mauritius, who began hybridizing in the 1820s and distributed his results in England through Robert Barclay of Bury Hill in Surrey. It is the present century, however, that has seen the greatest popularity of hibiscus breeding, with Hawaii and Southeast Asia the principal breeding centres.

Watercolour on paper, 60 × 47 cm
PROVENANCE: Reginald Cory bequest, 1936

PLATE 19

William Curtis (1746–1799)
Iris 'Vandewill'   Iridaceae

William Curtis was born in Alton, Hampshire, in 1746. From 1772 to 1777 he worked at the Chelsea Physic Garden; in 1777 he established his own botanic garden in Lambeth Marsh. In 1789 he started a nursery in Brompton, which moved to Chelsea after his death in 1799 and continued to trade until 1823. His most important works were the *Flora Londinensis* (1777–98), and the *Botanical Magazine*, which he started in 1787, and which continued under that title until 1983, to be succeeded by *The Kew Magazine*.

The Society possesses three coloured drawings of irises by Curtis, which were received as part of the bequest of E. A. Bowles in 1954. The drawing shown here is signed, but the iris is not named; but it is recognizably the same iris depicted on another drawing and labelled 'Vandewill' (on the verso of this drawing is a partially coloured pencil sketch of the same plant). The American Iris Society's checklist records its earliest reference to 'Vandewill' in 1839, when it was listed in the *Revue Horticole* among the plants in the late Nicolas Lemon's garden; its previous history is so far undocumented. It is regarded as synonymous with Späth's *Iris laevigata* var *belgica*. Tall bearded irises date back to the seventeenth century, but the earliest cultivars have disappeared.

Watercolour on card, 27 × 17.5 cm
PROVENANCE: E. A. Bowles bequest, 1954

PLATE 20

John Curtis (1791–1862)
Delphinium sp.   Ranunculaceae

After the death of William Curtis, his *Botanical Magazine* remained in the hands of his family until the 1840s. In 1818, John Curtis became its principal artist, and by 1832 contributed over 400 plates. He had been born in Norwich in 1791, and was no relation of William Curtis; he had already worked as an engraver for the Horticultural Society's *Transactions*. He is remembered primarily as an entomologist, although his multi-volume *British Entomology* (1824–39) includes illustrations of plants as well as insects. Between 1819 and 1832 he contributed over 400 plates to the *Botanical Magazine*. The Society's Library possesses several of his drawings, derived from two sources: some as part of a collection of *Botanical Magazine* drawings purchased in 1946 from Mr A. B. Burleigh of York, a descendant of Samuel Curtis, and others, of which this previously unpublished drawing is one, from the bequest of E. A. Bowles in 1954.

Some delphiniums were being grown in the sixteenth century, but their popularity later waned, and it was not until the last third of the nineteenth century that larkspurs and rockets returned to favour as old-fashioned plants, suitable for planting in historically revivalist gardens. James Kelway's hybrids accordingly found a ready market in the 1880s, and became the ancestors of the modern border delphiniums. In 1928 the British Delphinium Society was formed. The Royal Horticultural Society published a Checklist of delphinium names in 1949, and in 1966 became the International Registration Authority for the genus.

Watercolour on paper, 24.5 × 19.5 cm
PROVENANCE: E. A. Bowles bequest, 1954

J. Curtis del July 1823.

del Nov. 1818.

PLATE 21

John Curtis (1791–1862)
Chrysanthemum (*Dendranthema indica*)
Compositae

This is John Curtis's original drawing for plate 2042 of the *Botanical Magazine*, published in 1819 as *Chrysanthemum indicum*. All the early chrysanthemum introductions were of specimens taken from gardens, not from the wild. In 1903, Sir Joseph Hooker, remarking on the resulting confusion, decided that this drawing represented *C. morifolium* instead; but this species did not stand the test of time, and eventually was renamed *C. × grandiflorum*, an umbrella name for cultivated varieties raised in China from *C. indicum*.

Chrysanthemums were, at the time of the drawing, a recent introduction; it was in the 1790s that Colville's nursery in Chelsea staged the first good display. Twelve known varieties were named in the article accompanying this illustration, differentiated purely by the colour of their flowers. The specimen from which Curtis made his drawing was provided by Joseph Sabine, the Secretary of the Horticultural Society.

The genus *Chrysanthemum* was always something of a grab-bag, and suggestions have repeatedly been made about splitting it up. In 1855 Charles-Robert-Alexandre Des Moulins proposed that *C. indicum* was so different from Linnaeus's genus that it should be transferred into a new genus, *Dendranthema*; his proposal was adopted in Candolle's *Prodromus*, but rejected by Bentham and Hooker. In recent years, however, the break-up of *Chrysanthemum* has been agreed by the botanical world, and the garden chrysanthemums have been reclassified as *Dendranthema*. This need not greatly affect the ordinary gardener, however, for the status of 'chrysanthemum' as an English name is unaffected.

Watercolour on paper, 25 × 19.5 cm
PROVENANCE: purchased in 1946 from A.B.Burleigh of York

PLATE 22

Sydenham Teast Edwards (1768–1819)
Narcissus bulbocodium

Amaryllidaceae

Sydenham Teast Edwards was trained as an artist by William Curtis,. and made most of the illustrations for Curtis's *Botanical Magazine* from 1787 until 1815, when he resigned to found his own rival publication, the *Botanical Register*, which was later edited by John Lindley.

Narcissus bulbocodium, the hoop petticoat daffodil, was comparatively ignored by gardeners until its sudden surge of popularity in recent years. This section of *Narcissus* has an extraordinarily convoluted nomenclatural history, and a wide range of species and subspecies used to be listed. Anyone who is interested should read the chapter on bulbocodiums in E.A. Bowles' *Handbook of Narcissus* (1934); but since these were based on observing garden specimens rather than wild populations, they have now been swept away. 'Anyone trying to classify the hoop petticoats', wrote John Blanchard in an article in the Society's yearbook *Daffodils 1981–2*, 'finds that they come in a bewildering range of shapes, sizes and colours. They vary not only from location to location, but also to a remarkable extent within a single population.'

Watercolour on card, 22.5 × 19.5 cm
PROVENANCE: E.A. Bowles bequest, 1954

PLATE 23

Sydenham Teast Edwards (1768–1819)
Dianthus caryophyllus 'Wheat-ear'   Caryophyllaceae

This is Edwards' original drawing for plate 1622 of the *Botanical Magazine*. The published text says that the specimen was provided by 'Mr. M'Kirk', but the pencil note on the drawing (dated 10 September 1813) attributes it to 'Mr Kirk' – presumably Joseph Kirke (*c.*1768–1864), of the Cromwell's Garden Nursery in Kensington.

The *Botanical Magazine* discussed the wheat-ear carnation as a curiosity: 'We do not find this singular monstrosity mentioned by Parkinson ...' Linnaeus had distinguished this form as the variety *imbricatum*; its complex flower arose from the scales at the base of the calyx multiplying, and Linnaeus could hardly recall another instance of this phenomenon. Later researchers found it to be much more common than he had thought, and Maxwell T. Masters coined the name 'calycanthemy' for it.

The cultivation of different forms of carnations and pinks was widespread from the Middle Ages on. By the eighteenth century there was a range of categories (flakes, bizarres, flames) which florists' societies would exhibit in competition. The popularity of exhibition carnations did not decline, as that of most other categories of florists' flowers did; the National Carnation and Picotee Society was founded in 1877, and eventually swallowed its rival, the British Carnation Society, in 1949. In 1966 the Royal Horticultural Society became the International Registration Authority for *Dianthus*; in the second edition of its *Checklist* (1983), some 27,000 recorded cultivar names were listed.

Watercolour on paper, 25 × 19.5 cm
PROVENANCE: purchased in 1946 from A. B. Burleigh of York

PLATE 24

James Sowerby (1757–1822)
Helianthus giganteus Compositae

James Sowerby worked both as an engraver (e.g. for Sibthorp's *Flora Graeca*), and as a botanical artist. In addition to making several plates for the *Botanical Magazine*, he published a four-volume work of *Coloured Figures of English Fungi* (1795–1815), and illustrated two major publications for Sir James Edward Smith: *English Botany* (1790–1814), and *Exotic Botany* (1804–5).

The drawing shown here depicts *Helianthus giganteus*, labelled 'Helianthus crinitus'. This plant was first grown in England in 1714 by the Duchess of Beaufort at Badminton. Sir John Hill reported in the mid-eighteenth century that 'This is the tallest of the Helianthus kind; it is now thirteen feet high with me: but the Flowers for such a stature are not large'. David Douglas reported that its tubers were edible, and for a while it was known as the 'Indian potato'.

Linnaeus based his system of classification on the numbering of a plant's sexual organs, and during the heyday of his system, in the later eighteenth century, the flower became the most important feature to be depicted in a botanical illustration. Sowerby shows one way of adapting to this demand: the flower occupies the centre of the illustration, and is the only portion to be coloured, but the outline of the leaf is shown behind.

Watercolour on paper, 24 × 19.5 cm
PROVENANCE: E. A. Bowles bequest, 1954

Sowerby dd

PLATE 25

Ferdinand Bauer (1760–1826)
Isoplexis sceptrum   Scrophulariaceae

Ferdinand Lucas Bauer was born in Feldsberg, Austria, in 1760. Together with his brother Franz (1758–1840), he was employed by Nicolas Freiherr von Jacquin to illustrate his *Icones Plantarum Rariorum* in the 1780s. In 1784 he accompanied the English botanist John Sibthorp to Greece, and subsequently drew the illustrations for Sibthorp's posthumous *Flora Graeca*. From 1801 to 1805 he was botanical artist on Matthew Flinders' expedition to Australia; Flinders named Cape Bauer in his honour. Some of his drawings were published in 1813 as *Illustrationes Florae Novae Hollandiae*, though most remained unpublished until W.T.Stearn's edition of *The Australian Flower Paintings of Ferdinand Bauer* in 1976. Among his other works were the plates for Lambert's *Genus Pinus*.

In 1821 John Lindley published his monograph on foxgloves, *Digitalium Monographia*. Of the twenty-seven illustrations, twenty-one were made by Ferdinand Bauer, one by Franz Bauer, and five by Lindley himself. Shown here is Ferdinand Bauer's drawing of *Digitalis sceptrum*. By 1835 Lindley had reconsidered his classification, and proposed removing some *Digitalis* species into the new genus *Isoplexis*; this became one of them. It had been discovered in Madeira by Francis Masson, who sent it to Kew in 1777; it was introduced into cultivation by the Hammersmith nursery of Lee and Kennedy.

Watercolour on paper, 48.5 × 34.5 cm
PROVENANCE: Lindley library purchase, 1866

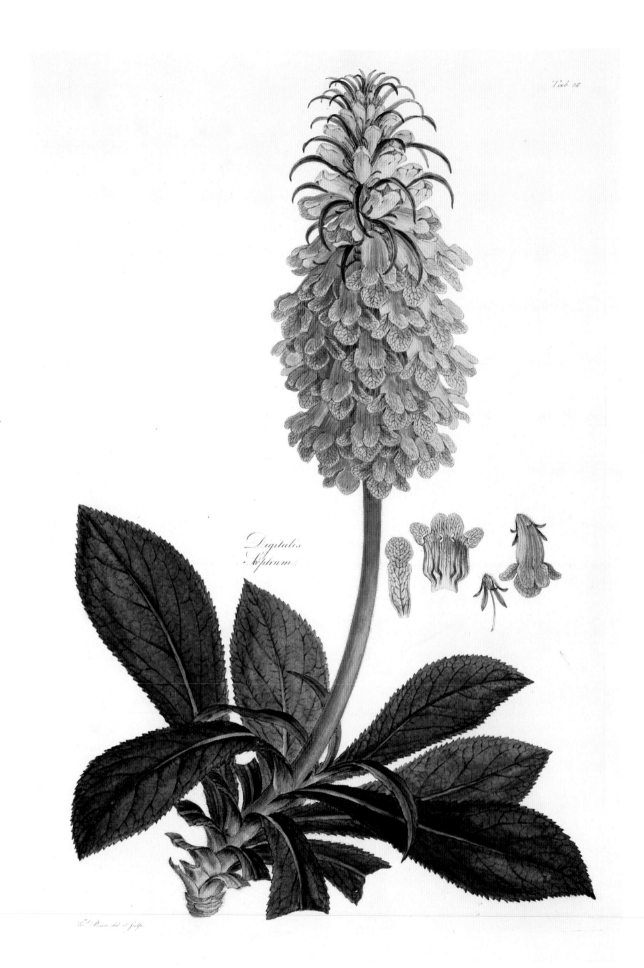

Tab. 98

*Digitalis
Sceptrum*

PLATE 26

Ferdinand Bauer (1760–1826)
Digitalis dubia   Scrophulariaceae

This drawing for Lindley's *Digitalium Monographia* depicts *Digitalis dubia*, known to Lindley as *Digitalis minor*. A native of Spain, it was introduced into England in 1789 by John Hunneman (*c*.1760–1839), a London bookseller and plant agent who catered to an international clientele. John Claudius Loudon, the founder of the *Gardener's Magazine*, wrote of him in 1829: 'There is not a botanist or reading gardener [Loudon never lost an opportunity to campaign for better education and improved literacy for gardeners] on the Continent or in this country to whom the name of Hunneman is not familiar; and by far the greater number of the former are under personal obligations to him, for transmitting them seeds, specimens, or books.'

Although *Digitalis* was much experimented upon by early hybridists in the nineteenth century, the results of their work were of scientific rather than garden interest. In the last third of the century, foxgloves enjoyed a revival of favour as 'old-fashioned flowers', and it was only then that nurserymen turned their attention to hybridization. The Veitch nurseries launched their Purpureo-grandiflora strain in the 1890s; William Wilks, the Secretary of the Royal Horticultural Society, bred the Shirley strain about the same time; further strains have appeared on the market in the succeeding century, through firms such as Suttons and Dobbies.

Watercolour on paper, 48.5 × 34.5 cm
PROVENANCE: Lindley library purchase, 1866

Digitalis
minor.

PLATE 27

William Hooker (1779–1832)
Apricot (*Prunus armenaica*) 'Breda'   Rosaceae

William Hooker was born in London in 1779, and studied under Franz Bauer. Between 1805 and 1808 he published the *Paradisus Londinensis*, with text by Richard Anthony Salisbury, a founder member of the Horticultural Society. It may have been through this connection that Hooker was employed by the Society as illustrator of its *Transactions*, and later as its fruit painter. During this period he also published his *Pomona Londinensis*. After 1820, however, ill health slowed his production, and in 1822 he stopped painting, probably as the result of a stroke. He died in 1832, and is commemorated in the colour name 'Hooker green'.

Native to Asia, apricots had been introduced to Europe by the time of Pliny in the first century AD, and to England by the time of William Turner in the mid sixteenth; the word 'apricot' is an eighteenth-century revision of the earlier 'apricock'. By 1688, Leonard Meager listed seven varieties; some seventy or eighty varieties were listed when the Horticultural Society began trying to sort out their nomenclature, but Robert Thompson, the Society's fruit specialist, reduced these to seventeen distinct forms. Robert Hogg described forty-nine sorts in his *Fruit Manual* (1884 edition), most of the newer additions being French in origin. Since then, with the decline of the kitchen garden with special glasshouses, the cultivation of apricots in England has declined drastically.

Shown here is Hooker's drawing of the Breda apricot. Richard Weston listed it in 1777; but the nomenclature of this variety is confused, and at least two other apricots have been confused with this one. The Breda is still extant today.

Watercolour on paper, 46 × 36.5 cm
PROVENANCE: commissioned by the Horticultural Society, 1819; sold 1859; re-purchased from John Napier of Leighton, 1926

W Stocker de
1819

The Breda Apricot

PLATE 28

Charles John Robertson (*fl.* 1820s)
Pear (*Pyrus communis*) 'Napoleon'   Rosaceae

On William Hooker's retirement in 1820, the Society hired Charles John Robertson to continue the work of depicting specimens of fruit exhibited at its meetings. Nothing is known about Robertson's life and career apart from the twenty-eight drawings of fruits in the Hooker series, made between 1820 and 1825. Only one of them was published in the Society's *Transactions*.

The pear is native to Europe and northern Asia, though probably not to Britain; but since old pear trees are mentioned as boundary trees in the Domesday Book, it must have been introduced long before the Norman conquest. Several varieties were available by the thirteenth century. The production of perry began much later in this country than in France, and was not discussed until the late sixteenth century; but in the seventeenth, men like Samuel Hartlib and John Evelyn promoted it as a local industry. Both perry production – despite the recent vogue for Babycham – and the cultivation of dessert pears have declined in England during the present century.

The pear shown here was raised in what is now Belgium in 1808, where it was first known as 'Médaille', but soon renamed 'Napoleon'. It was sent to the Horticultural Society in 1816 by the great Brussels fruit grower Jean Baptiste van Mons, and was quickly accepted as a valuable winter pear (fruiting in November and December). Robert Hogg, in his *Fruit Manual*, listed fourteen names under which it was known. It has disappeared from commerce in this country during the present century.

Watercolour on paper, 46 × 37.5 cm
PROVENANCE: commissioned by the Horticultural Society, 1820; sold 1859; re-purchased from John Napier of Leighton, 1926

Carolus Joannes Robertson Pinxit
Octob. 1820

Napoleon Pear.

PLATE 29

Barbara Cotton (*fl.* 1810s–1820s)
Peach (*Prunus persica*) 'Late Admirable'   Rosaceae

Barbara Cotton was the third of the Horticultural Society's fruit painters; hired in 1822, she made eight drawings in that year. Two of her plates were reproduced in the Society's *Transactions*. The drawings of waterlilies she made for the Society, unlike her fruit drawings, have not been recovered since the 1859 sale. Apart from that, little is known of her career; she lived in Newport Pagnell and exhibited at the Royal Academy between 1815 and 1822.

Peaches have been grown in China for millennia, and had spread to Greece by the fourth century BC. Peach stones have been found in one Roman excavation in this country, but it is not until the thirteenth century that there is documentary evidence for their cultivation here. From the sixteenth century onward, many new varieties were imported, until in 1826 the Horticultural Society could list 224 sorts, to be reduced to 183 as Robert Thompson sorted out their confused nomenclature. English gardeners and nurserymen bred many new varieties in the nineteenth century, some of which are still extant, but peach growing in recent times has suffered from the decline of the kitchen garden with its specialized peach houses. Despite Justin Brooke's mid-century attempts to promote open-air peach orchards, peach-growing in England is now confined to some two dozen varieties.

'Late Admirable', the peach shown here, was also known as 'Judd's Melting' and 'Motteux's Seedling'. E. A. Bunyard thought it was an old French variety dating back to the seventeenth century, but acknowledged there was confusion over its name.

Watercolour on paper, 46 × 37.5 cm
PROVENANCE: commissioned by the Horticultural Society, 1822; sold 1859; re-purchased from John Napier of Leighton, 1926

Barbara Cotton Pinx.t 1822

PLATE 30

Augusta Innes Withers (*c.*1793–1864)
Gooseberry (*Ribes uva-crispa*)   Grossulariaceae
'Crompton's Sheba Queen'

Augusta Innes Baker was born about 1793; her known works are signed by her married name of Withers. After working for the Horticultural Society in the 1820s, she was appointed flower painter to Queen Adelaide in 1830, and later became Flower and Fruit Painter in Ordinary to Queen Victoria. She died in 1864.

This gooseberry drawing, labelled 'Compton's Sheba Queen', was made for the Society in 1825. A slightly altered verision of this drawing was published in 1828 as the twelfth engraving in John Lindley's *Pomological Magazine*, under the name of 'Crompton's Sheba Queen': 'The name of the individual attached to the variety as its original grower, is erroneously spelled Compton in most of the Sale Catalogues of Fruits.'

Crompton was a Lancashire fruit grower, about whom no biographical information has yet surfaced. His gooseberry 'Sheba Queen' survived into the late nineteenth century, and was described in the last (1884) edition of Robert Hogg's *Fruit Manual*. No reference to it in the twentieth century has yet been traced.

The engraving published in the *Pomological Magazine* would strike most viewers as an excellent illustration, until they saw the original. An engraving can furnish no highlight brighter than the white of the page on which it is printed, whereas in the original painting, the translucence of the gooseberry skin is rendered in a manner probably not equalled by any other artist. This effect may have been achieved by using a white enamel undercoat.

Watercolour on paper, 46 × 36.5 cm
PROVENANCE: commissioned by the Horticultural Society, 1825; sold 1859; re-purchased from John Napier of Leighton, 1926

PLATE 31

John Lindley (1799–1865)
Rosa pimpinellifolia   Rosaceae

John Lindley was born near Norwich in 1799, the son of a nurseryman. His first book was *Rosarum Monographia* in 1820, and this drawing was made the following year. In 1822 he became Clerk of the Horticultural Society's garden at Chiswick, and spent the rest of his life with the Society, becoming Assistant Secretary in 1827 and Secretary in 1858; meanwhile, he was also serving as Professor of Botany at University College, Curator of the Chelsea Physic Garden, and editor of the *Gardeners' Chronicle* and the *Botanical Register*. It was his Library, purchased by the Society in 1866, that formed the nucleus of the present Lindley Library.

This drawing is labelled in pencil, 'Rosa spinosissima var pallida, Sabine MSS'. Joseph Sabine, the Horticultural Society's Secretary, had already noted in 1820 how confused the nomenclature of the Scotch roses was, and by the last edition of W.J.Bean's *Trees and Shrubs Hardy in the British Isles* the plant had been assigned to the species *pimpinellifolia*, on the grounds that Linnaeus's name *spinosissima* belonged to a species he had already described as *Rosa majalis*, and that his first published description that definitely applied to the Scotch rose called it *pimpinellifolia*.

Watercolour on paper, 46 × 36.5 cm
PROVENANCE: sold with the Society's library, 1859; repurchased from John Napier of Leighton, 1926

PLATE 32

Edwin Dalton Smith (1800–*c*.1866)
Rosa rugosa   Rosaceae

This drawing is labelled, in pencil, 'New ferox'. Why new? Perhaps because a plant formerly known as *Rosa ferox* had been renamed by John Lindley as *Rosa biebersteinii*, in order to distinguish it from the plant illustrated in Mary Lawrance's *Collection of Roses* (1799) as *Rosa ferox*. This latter rose is actually *Rosa rugosa*, described by Thunberg in 1784, and native to China, Japan, Korea, and Siberia; its circumstances of introduction into England are uncertain. One naturally occurring variety from the Kamchatka peninsula, now known as *ventenatiana*, was named in 1800 from a specimen in the garden of the French gardener Jacques-Philippe-Martin Cels, and had probably been introduced about 1770, earlier than the true species; so it may also be that the label 'new ferox' is intended to distinguish this from the previously known form. This species has also been known as *Rosa horrida*.

Edwin Dalton Smith flourished from the 1820s to the 1840s; nothing is known about his early life. Between 1825 and 1836 he illustrated Maund's *Botanic Garden*, and made drawings for several of Robert Sweet's works in the 1820s. During that same decade he also drew roses and chrysanthemums for the Horticultural Society, which were sold in 1859; the rose drawings were recovered in the 1920s.

Watercolour on paper, 46 × 35.5 cm
PROVENANCE: sold with the Society's library, 1859; repurchased from John Napier of Leighton, 1926

PLATE 33

James Sillett (1764–1840)
Dahlia 'The Sovereign'   Compositae

James Sillett was born in Norwich in 1764. He studied at the Royal Academy from 1787 to 1790, and exhibited there from 1796 to 1837; he died in 1840. Most of his career was passed in Norwich, where he is buried in the Rosary Cemetery; in 1815 he was president of the Norwich Society of Artists. In 1826 he published a book on flower painting.

This coloured drawing of the 'Sovereign' dahlia was made in 1823 for Mr C. Middleton, who raised the plant, and who presented the drawing to the Horticultural Society in 1826. It does not appear to have been a successful cultivar; its name has not yet been traced in contemporary trade catalogues. A dahlia called 'Sovereign' received an Award of Merit in 1893, but it was a yellow.

The dahlia was first flowered in Europe by the Spanish botanist Cavanilles in 1791, but it was not until the end of the Napoleonic wars, and the introduction into commerce of the varieties that the Empress Josephine had raised at Malmaison, that it became a widely popular plant. By 1829, J.C. Loudon could report that it was the 'most fashionable flower in this country', and hundreds of named varieties were listed in the 1830s. The National Dahlia Society was founded in 1882, and in 1966, the Royal Horticultural Society became the International Registration Authority for dahlias.

Watercolour on paper, 33 × 26 cm
PROVENANCE: sold with the Society's library, 1859; date of recovery not recorded

PLATE 34

Pierre-Joseph-François Turpin (1775–1840)
Pulsatilla vulgaris   Ranunculaceae

Pierre-Joseph-François Turpin (1775–1840) was one of the greatest botanical artists of his age; some, including the present author, value him more highly than his rival Redouté. Among the works which he illustrated were the 'Nouveau Duhamel' (*Traité des Arbres Fruitiers*, 1807–35), Humboldt and Bonpland's *Voyage aux Regions Equinoctiales*, Chaumeton's *Flore Médicale*, and La Billardière's *Sertum Austro-Caledonicum*, the original drawings for which are in the Society's library.

The drawing shown here is selected from an album of twenty-five drawings on vellum. As with many of Turpin's drawings, the details are so fine that he must have used a brush with a single hair.

Linnaeus absorbed Tournefort's genus *Pulsatilla* into *Anemone*, so the plant depicted here was long known as *Anemone pulsatilla*. Adanson proposed reinstating *Pulsatilla* as a separate genus in the 1770s, but it was only in the wake of Ernst Huth's monograph of 1897 that botanists gradually adopted the proposal. (Meanwhile, it remained familiar to the general public under its vernacular name of Pasque flower.) William Turner reported in the 1570s that it was common in the wild. In 1883, William Robinson noted the existence of 'several varieties, including red, lilac, and white kinds, but these are now rare'.

Ink, gold and watercolour on vellum, 30.5 × 21.5 cm
PROVENANCE: Reginald Cory bequest, 1936

1   4   3   2   5   6

PLATE 35

Pierre-Joseph-François Turpin (1775–1840)
Types of inflorescence

The *Leçons de Flore*, by J.L.M.Poiret (1755–1834) and P.J.F.Turpin (1775–1840), was first published in 1819–20 as a concluding part to F.P. Chaumeton's huge work, the *Flore Médicale*; it was then published separately. Two special copies were made for European monarchs, one for Louis XVIII and one for Franz II of Austria; they were printed on vellum, and the plates, instead of being printed and coloured, were individually re-drawn by Turpin. The copy made for Louis XVIII is now in the Society's Library. It is belived to have been brought to this country by Napoleon III, and then presented to his neighbour at Chislehurst, Sir Edward Henry Scott of Sundridge Park, whose bookplate appears inside. It presumably entered the market when the Sundridge estate was broken up early in this century; Cory bought it in 1931.

The text is a treatise on plant morphology; the illustrations proceed through types of roots, stems, and leaves to flowers and seeds, then illustrate competing systems of classification, and lastly depict specimens representing the different families in Jussieu's system. Shown here is plate 20, 'hermaphrodite flowers: dicotyledons', and the numbered flowers are: 1. *Asclepias*, 2. *Rosa*, 3. *Alcea*, 4. *Colutea*, 5. *Cassia*, 6. *Anethum*, 7. *Heracleum*, 8. *Delphinium*, 9. *Rosa* again, and 10. *Linaria*.

Watercolour and gold on vellum, 22.8 × 30.5 cm
PROVENANCE: Reginald Cory bequest, 1936

PLATE 36

Pancrace Bessa (1772–1830)
Cestrum macrophyllum   Solanaceae

Pancrace Bessa was born in 1772, and was a pupil of Spaendonck and Redouté; he died in 1830. Between 1810 and 1827, he produced 527 illustrations for a work entitled *Herbier Général de l'Amateur*, with text by Mordant de Launay; this work was republished in Brussels between 1828 and 1835, rearranged and with a new text by P. A. J. Drapiez, under the title *Herbier de l'Amateur des Fleurs*. The drawings were given by Charles X to the Duchesse de Berry, to whom Bessa had given painting lessons; she left them to her sister, the Empress of Brazil; in 1947 the collection was dispersed at auction. The Society bought four of the drawings in 1970.

Cestrum macrophyllum, according to Mordaunt de Launay, was introduced from Puerto Rico, and had been grown by the great gardener Jacques-Philippe-Martin Cels since the 1790s. In the revised version, Drapiez said that these plants 'contribute greatly to the adornment of our temperate houses during most of the winter, and to the decoration of our gardens, when the season allows them to be placed out'. They later fell from fashion, however, as insufficiently ornamental; when Thomas Baines wrote his *Greenhouse and Stove Plants* in 1885, he discussed only one cestrum, *C. aurantiacum*, and said dismissively, 'This belongs to a family of plants that are not particularly attractive'.

Watercolour on paper, 26.5 × 21 cm
PROVENANCE: Purchased at Christie's, 1970

PLATE 37

Anonymous
Paeonia suffruticosa, double purple form   Paeoniaceae

John Reeves (1774–1856) was an inspector of tea at Canton and Macao for the East India Company. In 1817, he offered to send plants and drawings to the Society. By 1819 the Society had received 130 drawings, and others continued to arrive during the 1820s, making a total of 755. By the time they were sold in 1859, they had been divided into two sets, one consisting of three folio volumes containing 127 drawings of camellias, chrysanthemums, and peonies, the remaining 628 in a group of five volumes. The five-volume set came back to the Society through Reginald Cory's bequest; the three-volume set was repurchased in 1953 for £400, a sum nominally greater than the entire sale price of the Library in 1859.

The tree peony or moutan (or *mudan*, in the pinyin transliteration that is now recommended) has been grown in China for centuries, and is almost unknown in the wild. The first moutan arrived in Europe in 1789, planted at Kew by Sir Joseph Banks; by 1826, Joseph Sabine could list nine sorts in cultivation, and refer to this drawing as evidence for what still awaited the collector. The next great wave of introductions came at mid-century, through the agency of Robert Fortune; James Bateman, in the 1850s, experimentally planted an area of his garden at Biddulph Grange, Staffordshire, with moutans. Further species were introduced in the early twentieth century, and Reginald Farrer became the first European to see a wild form of the plant, near the Tibetan border.

Watercolour on paper, 49 × 38.5 cm
PROVENANCE: gift of John Reeves, *c.*1820; sold 1859; repurchased from Heywood Hill, 1953

PLATE 38

Anonymous

Chrysanthemum (*Dendranthema indica*)   Compositae
'Purple Phoenix Tail'

In addition to drawings, John Reeves sent plants to the Horticultural Society. In 1821 he sent twelve Chinese garden chrysanthemums; by 1824 the Society's garden at Chiswick had twenty-seven cultivars, and forty-eight by 1826. In 1843 the Society sent Robert Fortune to collect plants in China; he returned with more new chrysanthemums, and later introduced the first Japanese cultivars in 1862.

This drawing is annotated in John Lindley's handwriting, 'Chrysanthemum indicum var. Purple Pheasants Tail'. This version of the name was supplied by Reeves, and was cited by Joseph Sabine in an 1824 article in the Horticultural Society's *Transactions* on the new Chinese chrysanthemums, in which he said, 'many of these [names] are curious and fanciful, and if they could be adopted, would afford a variation in our nomenclature, as well as relieve us from some perplexity in giving names derived from the colours of the blossoms, a difficulty which will increase upon us as the number of our collection increases'. Nonetheless, this name is a mistranslation of the Chinese characters on the lower right, which actually read 'Purple Phoenix Tail'.

Watercolour on paper, 49 × 36 cm
PROVENANCE: gift of John Reeves, *c.*1820; sold 1859; repurchased from Heywood Hill, 1953

紫鳳尾 Tsze.
Fung.
Ny.

Chrysanthemum Indicum. Var. Purple Pheasants Tail.

PLATE 39

Anonymous
Camellia hongkongensis   Theaceae

The first camellias reached England in the 1730s, but it was not until the
1790s that the introduction of different varieties began to pick up speed. By
the 1820s, nurserymen from England to Italy were breeding new forms, and
several beautifully illustrated camellia books were published in the second
quarter of the century. The taste of the period favoured camellias with
symmetrical and geometrically complex flowers, often as conservatory
plants; in the last quarter of the century, these began to fall out of fashion,
and were replaced by those camellias with larger, looser flowers, for use in
the woodland garden.

This camellia was discovered in Indochina in 1837 by Gaudichaud-
Beaupré; then about 1849 it was observed by John Eyre in Hong Kong.
At first it was taken to be the true type of *Camellia japonica*, but in 1859
Berthold Seemann distinguished it as a new species, which he called
C. hongkongensis. (Seemann mistakenly said that there were only three trees
of it still extant in Hong Kong, but there were in fact hundreds.) This camellia
was introduced into cultivation in 1874, and continues to be commercially
important today.

This painting comes from a collection of 100 coloured drawings by
unknown Chinese artists, bound in four volumes.

Watercolour on paper, 43 × 35 cm
PROVENANCE: Purchased, 1912

PLATE 40

Anonymous
Magnolia denudata   Magnoliaceae

The yulan has been grown in China for centuries as an ornamental tree. It was introduced into Europe by Sir Joseph Banks in 1780, and quickly became one of the most popular magnolias in English gardens. It is highly variable in cultivation, and has been much used in hybridization.

The nomenclatural history of this magnolia is complex. The drawing is labelled 'Magnolia chulan' (a misrendering of 'yulan'); a later hand has added *Magnolia conspicua*, a name coined by R. A. Salisbury in 1806, and the name it was known by in England during the nineteenth century. Under the principle of priority, however, the name *Magnolia denudata*, published by Desrousseaux in 1791, has been accepted as the legitimate name since 1913, 'even though', as G. H. Johnstone said in his classic *Asiatic Magnolias in Cultivation* (1955), 'there are some ambiguities in his description, for instance he refers to the flowers as "rouges"'. In 1934 J. M. Dandy attempted to replace this with the name *Magnolia heptapeta*, on the grounds that Buc'hoz had illustrated this plant in 1779 under the name *Lassonia heptapeta*; Johnstone rejected this claim because Buc'hoz's illustration was incorrect, and Dandy eventually yielded. It is never safe to assume that a name is stable, however, and there has been a more recent attempt to re-assert Buc'hoz's name.

This illustration is reproduced from a folio of seventy-two coloured drawings of flowering plants, by an unnamed Japanese artist of the nineteenth century.

Watercolour on paper, 41.5 × 33 cm
PROVENANCE: Reginald Cory bequest, 1936

PLATE 41

Anonymous
Rhododendron indicum, red double form   Ericaceae

Linnaeus's genus *Azalea* was absorbed into *Rhododendron* in 1834 by George Don, a former Horticultural Society plant collector. American azaleas had been trickling into England since the 1690s, but it was not until the arrival of *Azalea pontica* (later renamed *Rhododendron luteum*) in the 1790s that the possibilities of cultivated varieties began to be realized. The cultivars collectively known as Ghent azaleas began to be bred after the Napoleonic wars, and quickly became favourite plants for the greenhouse and conservatory. Japanese azaleas played a part in breeding from the 1830s. This drawing, labelled 'Azalea indica flore pleno', is taken from a folio of seventy-two coloured drawings of flowering plants, by an unnamed Japanese artist of the nineteenth century.

The great career of the rhododendron in this country began in the nineteenth century. Some species, like *Rhododendron hirsutum* and *R. ponticum*, were introduced during the seventeenth and eighteenth centuries, but it was only about 1840, when gardeners discovered that *R. ponticum* could seed itself in England, that their landscape use began to be exploited. In 1847–51, the first seed of Himalayan species was sent back from Sikkim by Joseph Hooker, and the breeding of hardy hybrids had begun within a decade. By the turn of the century, attention was turning to China, and collectors such as George Forrest, E. H. Wilson, and Frank Kingdon-Ward sent back hundreds of species, which were seized upon by hybridizers. Today there are close to 5000 rhododendron cultivars, not including 6000 azaleas; the Royal Horticultural Society is the International Registration Authority.

Watercolour on paper, 41.5 × 32 cm
PROVENANCE: Reginald Cory bequest, 1936

Azalea Indica

Fiore pleno

PLATE 42

Valentine Bartholomew (1799–1879)
Hyacinthus orientalis cultivars   Liliaceae

Hyacinths were introduced into Europe from Turkey in the sixteenth century, and quickly gave rise to a multitude of varieties. At their peak of favour, in the mid-eighteenth century, the Dutch were supposed to grow some 2000 cultivars. Since then their popularity has declined. Between 1864 and 1898, fifty-seven hyacinth cultivars received First Class Certificates or Awards of Merit; none has done so since, and of those fifty-seven only two are still available.

Valentine Bartholomew was Flower Painter in Ordinary to the Duchess of Kent, and then to Queen Victoria. In 1821–22 he collaborated with the pioneering lithographer Charles Hullmandel in publishing *A Selection of Flowers adapted Principally for Students*, containing twenty-four hand-coloured plates (although the Lindley Library's copy has an additional twelve, not recorded in the usual bibliographies).

Bartholomew's second wife, Ann Charlotte (1800–62), was a poet and playwright, and is buried with him in Highgate Cemetery. In her 1840 volume *The Song of Azrael*, she included a poem entitled 'On seeing a fly, one autumnal day, on Mr. Bartholomew's beautiful picture of "The Gardener's Shed"':

> Thou art mistaken, foolish fly!
>   There dost thou puzzled stand;
> Those rose-buds, that with nature vie,
>   Are *not* by nature's hand.
>
> How wondrous is that painter's power!
>   To make thee thus believe
> Thou flutterest round a living flower: –
>   Yet man it might deceive!
>
> Poor insect, with thy feeble wing,
>   Still on that leaf remain;
> *Such flowers*, for a time must bring
>   Thy summer back again.

Watercolour on paper, 41 × 25.5 cm
PROVENANCE: Reginald Cory bequest, 1936

V. Bartholomew.
1851.

PLATE 43

Sir Charles James Fox Bunbury (1809–1886)
Protea repens   Proteaceae

The genus *Protea* is African, and largely concentrated in South Africa. The protea illustrated here is the most widely distributed of the Cape species; Bunbury, the artist, described it forming 'a thick belt of shrubbery along the eastern flanks of the Devil's and Table mountains'. It was the first protea to be cultivated in Europe; Francis Masson brought it to Kew in 1774, where it first flowered some six years later.

Sir Charles James Fox Bunbury collected plants in South Africa, South America, Madeira, and the Canary Islands. As a brother-in-law of Charles Lyell, he was also interested in geology, and at one time planned a revision of Lindley and Hutton's work on fossil botany. The drawing shown here, labelled 'Protea mellifera (sugar bush), Cape, July 1838', is a specimen from a folio album of drawings of plants and animals, made by Bunbury while in South America and South Africa.

Bunbury knew the plant under Carl Thunberg's now obsolete name, *Protea mellifera* (honey-bearing). In his *Botanical Fragments* (1883), Bunbury described the flower-heads as 'overflowing with honey, which attracts swarms of bees and beetles of all kinds. The quantity of honey in these flowers, when they first expand, is so great that by merely inverting them, one can pour it out as from a cup.' This was long used for making a local syrup and cough medicine called bossiestroop, until production ceased about 1900.

Watercolour on paper, 28 × 22 cm
PROVENANCE: Deposited on loan from the Bunbury family, 1965

PLATE 44

Walter Hood Fitch (1817–1892)
Passiflora antioquiensis   Passifloraceae

Walter Hood Fitch made over 2700 plates for *Curtis's Botanical Magazine* from 1834 to 1877, as well as illustrations for *Icones Plantarum*, Hooker's *Rhododendrons of Sikkim-Himalaya*, and Elwes' *Monograph of the Genus Lilium*. He frequently (in the last-named case, for instance) drew directly onto the lithographic stone, so that no separate original drawings survive for some of his works. He produced so much work that his reputation for facility came to be held against him, but his drawings always illustrated the accompanying description with great accuracy.

The drawing shown here is a case in point. Fitch made two versions of this drawing in the same year (1866), one of which appeared in the April issue of the *Botanical Magazine*, and the one shown here in the August issue of the *Florist and Pomologist*. It depicts the red banana passion flower, *Tacsonia van-volxemii* (now *Passiflora antioquiensis*). The description in the *Florist and Pomologist* concentrated on the flowers and on cultivation; accordingly, Fitch copied his earlier drawing (so it appears inverted from the *Botanical Magazine* version), and left out structural details that were not mentioned in the later article.

Watercolour on paper, 28 × 19.5 cm
PROVENANCE: Reginald Cory bequest, 1936

PLATE 45

Walter Hood Fitch (1817–1892)
Chrysanthemum (*Dendranthema indica*)   Compositae
'Gloria Mundi'

In 1832 the first chrysanthemum seed was raised in England, and the breeding of new varieties for garden and exhibition began. The first chrysanthemum show was held in Norwich in 1843. In 1846, the Stoke Newington Chrysanthemum Society was founded; in 1884, with its third change of name, it became the National Chrysanthemum Society.

One of the first great names among English chrysanthemum breeders was John Salter (1798–1874). In 1838, deciding to devote himself to chrysanthemums, he moved to Versailles, where he thought the climate was better for them, and produced his first seedlings in 1844. When revolution broke out in 1848, he moved back to England, setting up the Versailles Nursery at Hammersmith, and becoming the authority on the flower for his generation. He published a book on the subject in 1865.

He exhibited his new cultivar 'Gloria Mundi' in 1865, when it received a First Class Certificate. 'A remarkably fine variety', reported the Floral Committee, 'with deep incurved blossoms, of a rich intense golden yellow. Its brilliant colour and fine form were greatly admired.' The *Florist and Pomologist* described it as 'the finest flower of the year … a model of form'. In 1890, it was still listed as one of the best incurved forms, although by a minority of votes that indicated its popularity was fading.

Watercolour on paper, 28 × 18.8 cm
PROVENANCE: Reginald Cory bequest, 1936

PLATE 46

Henry George Moon (1857–1905)
Calochortus clavatus and *C. nitidus*

Liliaceae

Henry George Moon was born in London in 1857. In 1880 he joined William Robinson's magazine *The Garden* as an artist, and drew plants for it for much of his career. His most celebrated illustrations were made for Frederick Sander's great orchid book *Reichenbachia*, and for another of William Robinson's magazines, *Flora and Sylva*, which was started in 1903 and discontinued on Moon's death in 1905.

Robinson praised Moon for his accuracy in depicting the habit of plants: 'Before the days of the *Garden* plates it was a common practice to exaggerate the drawing of flowers. There was a false florist's ideal set up to which all had to conform: a circle and a large Cauliflower being the accepted models ... we began with the determination to draw things as they drew themselves, and never deviated from it ...' What Robinson was referring to here was not the standard of botanical illustration in publications like *Curtis's Botanical Magazine*, but rather the illustrations of florists' flowers in magazines like the *Floricultural Cabinet*. (See Fitch's chrysanthemum in plate 45 for an example.) Moon's drawings were intended to help the gardener determine what effect plants would have in cultivation, rather than to help the botanist in identification. His work had a great impact on British plant drawing in the early twentieth century; artists such as E. A. Bowles, Dorothy Martin, and Lilian Snelling in her early days all show his influence.

Shown here is Moon's original drawing for a plate in the first volume of *Flora and Sylva*, showing two species of the American genus *Calochortus*: *C. clavatus* (introduced into commerce by the American grower Carl Purdy in the 1890s) and *C. nitidus* (introduced by David Douglas in 1826).

Watercolour on paper, 38 × 27 cm
PROVENANCE: Purchased, 1991

PLATE 47

(?) Arthur Perry (*fl.* 1890–1910)
Narcissus 'Elvira'   Amaryllidaceae

Daffodils, having been comparatively neglected as garden plants, returned to favour in the last third of the nineteenth century as old-fashioned flowers. Peter Barr, the Covent Garden bulb merchant, tried to recover all the varieties of daffodil listed in Gerard and Parkinson. A great wave of daffodil breeding followed. In the Netherlands, the nurseryman R. A. van der Schoot began crossing *Narcissus poeticus* with *N. tazetta* in the 1880s; the resulting Poetaz hybrids began to appear on the market after the turn of the century. One of his first was 'Elvira', which received an Award of Merit in 1904, and which is still extant, though no longer commercially available.

This is one of a series of 172 drawings made for William Fowler Mountford Copeland (1872–1953), of the Staffordshire pottery firm W. T. Copeland and Sons, who appears to have borrowed the firm's best artists to depict the daffodils he was growing, and between 1895 and 1907 compiled a collection of 172 daffodil portraits in gouache. The artist may have been one Arthur Perry, but three different hands can be distinguished in the collection.

Gouache on paper, 27.5 × 19 cm
PROVENANCE: Purchased from W. F. M. Copeland, 1935

Elvira

PLATE 48

(?) Miss Williamson (*fl.* 1900s)
Iris laevigata var *albopurpurea*   Iridaceae

The name *Iris albopurpurea*, which appears on this drawing, was coined by J. G. Baker in 1896 to describe a Japanese iris which had been grown at Kew that year, flowering by the side of the lake in front of the Palm House from the middle to the end of June. William Rickatson Dykes, remarking in his *Genus Iris* that 'Owing to the fact that *I. kaempferi* and *I. laevigata* have been looked upon as synonymous, the literature has become very confused', reclassified it as *Iris laevigata* var *albopurpurea*, a quasi-albino form of *laevigata*.

This picture, dated 29 June 1905, is taken from an album of thirty-five coloured drawings of irises, made for Ellen Willmott during the years 1904–08, apparently from varieties flowering in her garden at Warley Place, near Brentwood in Essex. Willmott (1858–1934) was described by Gertrude Jekyll as 'the greatest of all living women gardeners'; her garden, celebrated for its collection of 100,000 species and varieties of plants, fell into dereliction after her death, and is now maintained as a country park.

The album was bequeathed to the Society by Miss Willmott. No artist's name is recorded on the paintings, but they are said to have been the work of a Miss Williamson. This might be Ella Williamson, whose address during the Edwardian period was given as Paris, but who exhibited at the Society of Women Artists.

Watercolour on paper, 31.5 × 23.5 cm
PROVENANCE: Bequest of Ellen Willmott, 1935

Iris albopurpurea

Flowering state 29th June 1905

actual size

PLATE 49

(?) Miss Williamson (*fl.* 1900s)
Iris fulva

Iridaceae

Friedrich Pursh named this *Iris cuprea* in 1814, unaware that John Ker-Gawler had already named it *Iris fulva* in the *Botanical Magazine* two years earlier. This was unfortunate, for *cuprea* drew attention to its most significant feature: its bright copper colouring, which prompted numerous attempts in the late nineteenth century to hybridize it. W. R. Dykes finally succeeded in 1910, with his hybrid *Iris × fulvala*.

This iris is native to the lower Mississippi valley, and attempts by British gardeners to imitate its native growing conditions may have helped to delay its popularity. Dykes wrote in *The Genus Iris*: 'The leaves of this Iris show distinctly by their structure that it is an aquatic species, but it remains flowerless, if treated as such in England. It requires here a somewhat dry and warm position in rich light soil.'

The popularity of irises grew enormously in the late nineteenth century, and the Cambridge physiologist Sir Michael Foster planned to write the first survey of the genus. In the end, he passed this task to William Rickatson Dykes, who became the leading authority and the author of *The Genus Iris* (1913) and *Handbook of Garden Irises* (1924); in addition, his collected articles on the subject were posthumously published as *Dykes on Irises* (1932). He also served as Secretary of the Royal Horticultural Society until his death in a car crash in 1925.

Watercolour on paper, 31.5 × 23.5 cm
PROVENANCE: Bequest of Ellen Willmott, 1935

Iris Cuprea.
Flowering state 29ᵗʰ June 1905.
actual size.

PLATE 50

Alfred Parsons (1847–1920)
Rosa × alba 'Celeste'   Rosaceae

Alfred William Parsons (1847–1920) was a botanical artist and landscape gardener who spent much of his career in the artists' colony at Broadway, Worcestershire. He was President of the Royal Society of Water Colour Painters from 1914 until his death. In 1901 he began painting the roses in Ellen Willmott's garden at Warley Place for an intended publication. Although he had completed most of the paintings by 1906, the first part did not appear until 1910, and publication was not completed until 1914. *The Genus Rosa*, published under Ellen Willmott's name, with text by J. G. Baker and 132 illustrations by Parsons, sold only 260 copies. Willmott was heavily in debt during her last years, and sold her own copy of the work, with Parsons' original drawings bound in facing the printed plates; this copy is now in the Lindley Library. The published chromolithographs were greatly inferior to Parsons's drawings, and in 1987 a selection of the plates was newly reproduced from the originals under the title *A Garden of Roses*, with text by Graham Stuart Thomas.

Baker classified this rose as *Rosa alba* var *rubicunda*, and claimed that it was the *Rosa incarnata* of Parkinson and the 'Maiden's Blush' of Philip Miller. Graham Thomas, however, has identified it as 'Celeste', a cultivar of Dutch origin in the late eighteenth or early nineteenth century.

Watercolour on paper, 30.5 × 23.5 cm
PROVENANCE: Reginald Cory bequest, 1936

137

PLATE 51

Edward Augustus Bowles (1865–1954)
Anemone coronaria var *phoenicia*   Ranunculaceae

Edward Augustus Bowles was born in 1865, and died in 1954, having been active in the Royal Horticultural Society's affairs for fifty years. He served at one time or another on the Society's Council and several of its committees, serving as Chairman of three of them, in particular the Narcissus and Tulip Committee, which he chaired for over forty years, from 1911 to 1954. His bequest of his drawings and papers was received by the Society in 1954–5, and a Bowles Memorial Travel Scholarship was established in his name.

Bowles' garden at Myddelton House, Enfield, became famous as the result of his books *My Garden in Spring, My Garden in Summer*, and *My Garden in Autumn and Winter*. He also wrote *A Handbook of Narcissus* and *A Handbook of Crocus and Colchicum*. Left unfinished at his death was a project on *Anemone*, portions of which appeared as articles, one written in collaboration with W. T. Stearn, in the *Journal of the Royal Horticultural Society*.

Shown here is Bowles' coloured drawing of *Anemone coronaria* var *phoenicia*, made at Menton in 1928. In *My Garden in Spring* (1914), he reported that the '*coronarias* are never happy here [Myddelton] for long, and their cultivation must be pursued on the buy-and-die system, which I dislike as wasteful and unkind to the plants. But their price is so low that I occasionally invest in a hundred or so to get at least one season's fun out of them.'

Watercolour on board, 38 × 27 cm
PROVENANCE: E. A. Bowles bequest, 1954

Anemone coronaria
var. phoenicia
Mentone
9-III-1928

PLATE 52

Edward Augustus Bowles (1865–1954)
Galanthus 'Galatea'   Liliaceae

Galanthus nivalis first enters the literature as a garden plant, under the name 'bulbous violet'. The word 'snowdrop' did not enjoy widespread use until the later seventeenth century. The plant is probably not native, and was not observed in the wild until the 1770s, when William Withering found some near Malvern. In the nineteenth century, however, a spurious antiquity was constructed for the snowdrop, as in the lines:

> The Snowdrop in purest white arraie
> First rears her hedde on Candlemas daie,

allegedly from a medieval poem, though in fact written by Thomas Ignatius Forster in the 1820s. The great snowdrop fields seen in some English gardens are probably little more than a century old.

The great breeder of snowdrops was James Allen of Shepton Mallet, who was active from the 1880s until his death in 1906. Few of his cultivars survive; his collections were reduced by infestation before his death, and by mid-century, Bowles reported that Allen snowdrops were 'very scarce in English gardens, owing to their slow rate of increase'. 'Galatea' was one of Allen's most famous achievements, and is still commercially available. At the Snowdrop Conference of 1891, he described it as 'one of the giants of the family as to size of flower, but not in stature'.

E. A. Bowles never completed his intended book on *Galanthus*; Sir Frederick Stern incorporated his material in his *Snowdrops and Snowflakes* (1956). Among the papers Bowles bequeathed to the Royal Horticultural Society was a collection of 112 drawings of snowdrop varieties, from which this 1916 portrait is taken.

Watercolour on paper, 35.5 × 25 cm
PROVENANCE: E. A. Bowles bequest, 1954

PLATE 53

Lilian Snelling (1879–1972)
Paeonia bakeri   Paeoniaceae

Lilian Snelling spent most of her life in St Mary Cray, Kent, where she was born in 1879. During the Edwardian period she came under the patronage of H.J. Elwes, for whom she made many drawings of plants at his garden at Colesbourne, Gloucestershire. Elwes had previously been a patron of Walter Hood Fitch, who had illustrated his *Monograph of the Genus Lilium*, and Snelling was to follow in Fitch's footsteps both on lilies and on the *Botanical Magazine*.

From 1916 to 1921 she worked at the Edinburgh Botanic Garden, under the supervision of Sir Isaac Bayley Balfour. From 1922 to 1952 she illustrated *Curtis's Botanical Magazine*, which had recently been acquired by the Royal Horticultural Society; volume 169 was dedicated to her on her retirement. She also illustrated the *Supplement to Elwes' Monograph of the Genus Lilium*, and, for the Society, Sir Frederick Stern's *Study of the Genus Paeonia*. She was awarded the Victoria Medal of Honour in 1955.

Shown here is one of Snelling's illustrations for Stern's peony monograph. *Paeonia bakeri* is a plant known only from gardens; all specimens are descendants of a plant of unknown origin found in the Cambridge Botanic Garden and described by R. Irwin Lynch in 1890. Nothing corresponding to it has since been found in the wild. It appeared at a time of great interest in peony hybridization, when firms like Kelway in England and Lemoine in France were busy importing new species for breeding; Peter Barr introduced it into commerce in 1895.

Watercolour on paper, 35.5 × 25.5 cm
PROVENANCE: Commissioned by RHS

PLATE 54

Lilian Snelling (1879–1972)
Primula 'Marven'   Primulaceae

Auriculas (*Primula* × *auricula*), a continental introduction, were familiar in England by the late sixteenth century, and by the early nineteenth, Isaac Emmerton could list over ninety varieties available. They were one of the most popular of 'florists' flowers', grown for competition by societies of amateur breeders and exhibitors; English florists were particularly devoted to varieties with distinct edges. During the nineteenth century, the older florists' societies dwindled; despite the formation of the National Auricula Society in 1873, the popularity of the exhibition auricula declined to the point where only one English nursery (Douglas of Bookham) continued to grow the old stocks. In 1979 Brenda Hyatt staged an exhibit of these varieties at the Chelsea Flower Show, and since then their popularity has once again soared.

'Marven' was a cross between *Primula* × *venusta* (itself a cross between *P. auricula* and *P. carniolica*) and *P. marginata*. At the Primula Conference of 1913, Reginald Farrer held it up as an example of the 'skill and success' of the modern plant breeder. It had to wait for an Award of Merit until 1967, when it was exhibited by K. N. Dryden as a 'hardy flowering plant for the alpine house and rock-garden'. The fluctuating status of the auricula is illustrated by the fact that, while today 'Marven' is listed as an auricula hybrid, on its first appearance it was described as a marginata hybrid.

This drawing was made by Snelling for H. J. Elwes from a plant in his garden at Colesbourne, and is dated 1915.

Watercolour on paper, 35.5 × 25.5 cm
PROVENANCE: Snelling bequest

Ruth Colleridge rdfc. 1922.

PLATE 55

Ruth Collingridge (*fl.* 1910s–1920s)
Anemone 'Honorine Jobert'

Ranunculaceae

The plant known to gardeners as *Anemone japonica* (*A. hupehensis* var *japonica*) was introduced by Robert Fortune in 1844. It first flowered at the Horticultural Society's garden at Chiswick in 1845, and before the year's end John Lindley had it illustrated in the *Botanical Register*. By 1847, George Gordon, the superintendent at Chiswick, could report in the Society's *Journal* that it was hardy, 'one of the most desirable of herbaceous plants for autumn decoration', and that he hoped to produce 'new varieties … by hybridizing the Japan anemone with such kinds as the large white *Anemone vitifolia*, from the north of India …' The first such hybrid was exhibited in 1849.

'Honorine Jobert' was bred by a Verdun nurseryman in the late 1850s. It was first noticed in this country when it was exhibited by F. and A. Smith of Dulwich at a Royal Horticultural Society show in September 1863, and received a commendation as 'a good late-blooming hardy border plant'; the following month a specimen in better condition was exhibited by E. G. Henderson of the Pine-Apple Nursery, and this time it received a Second-class Certificate (the old title of the Award of Merit). It is still widely grown today.

Little is known about Ruth Collingridge; she lived in Brentwood, Essex, and exhibited at the Society of Women Artists and the Dudley Gallery between 1912 and 1921.

Pencil and watercolour on paper, 47.5 × 28 cm
PROVENANCE: Reginald Cory bequest, 1936

PLATE 56

Frances L. Bunyard (*fl.* 1920s–1930s)
Cherries (*Prunus avium*)
'Kentish Bigarreau' and 'Early Rivers'

Rosaceae

Cherries, according to Pliny, had been introduced to Britain by the Romans by the middle of the first century AD. Whether or not they lapsed from cultivation after the departure of the Romans and had to be reintroduced, they were a popular fruit during the Middle Ages; thirty varieties were available by the early seventeenth century, and 246 by the nineteenth. This number was augmented by breeding programmes during the past two centuries. Since the 1950s the acreage of cherry orchards in Britain has declined rapidly.

Of the varieties shown here, 'Kentish Bigarreau' (top), also known as 'Bigarreau Kentish' and 'Graffion', is the older, and was used by Thomas Andrew Knight in his cherry breeding in the early nineteenth century. At the beginning of this century, according to E. A. Bunyard, this was 'the variety most commonly grown in Kent'. 'Early Rivers', probably the most famous of all English cherries, first fruited in the nursery of Thomas Rivers, in Sawbridgeworth, Hertfordshire, in 1869; it was awarded a First Class Certificate in 1872, and is still available today.

The Bunyard family ran the Royal Nurseries at Allington, near Maidstone. George Bunyard was one of the most important fruit growers of the late nineteenth century; his son Edward Ashdown Bunyard, who succeeded him, introduced the 'Golden Delicious' apple into England. Frances Bunyard was Edward's sister. In 1937 she exhibited fruit drawings at the Society's show and was awarded a Silver Grenfell Medal; she gave the Society a collection of fifty-three coloured drawings of cherry varieties, of which these are specimens.

Watercolour on paper, average size 20 × 20 cm
PROVENANCE: Gift of the artist, *c.*1933

PLATE 57

Dorothy Martin (1882–1949)
Ilex aquifolium   Aquifoliaceae

Dorothy Burt Martin was born in 1882, and studied at Wolverhampton School of Art and the Royal College of Art. From 1916 to 1949 she served as art mistress at Roedean. She exhibited occasionally over the years, but concentrated her effort on a projected British flora, which she never completed. She found her specimens in the Sussex countryside, and in the Lake district, where the school was evacuated during World War II. After her death in 1949, her sister Margaret presented a collection of 305 paintings of British plants to the Society.

The common holly, *Ilex aquifolium*, grows everywhere in Britain except northeast Scotland. It is highly variable, in habit, leaf and fruit colour, and leaf shape; Elwes and Henry, in their *Trees of Great Britain and Ireland* (1913), listed forty-nine varieties, sixteen of which are still commercially available.

Enthusiasm for the holly could be said to have reached its peak in 1852, when the great gardener William Barron published a proposal for replacing hawthorn hedges all over the country by holly hedges, and thus turning the country into *The British Winter Garden* (the title of his book).

Watercolour on paper, 59 × 39 cm
PROVENANCE: Presented by the artist's sister, 1949

Aquifoliaceae

Ilex Aquifolium.
Holly.

(Evergreen Shrub or Tree)

PLATE 58

Dorothy Martin (1882–1949)
Viscum album   Viscaceae

The mistletoe was formerly known in Latin as *Lonicera*, a name which Linnaeus borrowed as a generic name for the honeysuckles (themselves formerly known as *Caprifolium*). Linnaeus' arbitrary redistribution of Latin names caused some angry comment in the mid eighteenth century, but has generally been overlooked in the interests of a standardized nomenclature.

Mistletoes are unusual in that they are flowering plants, yet have adopted a parasitic habit. *Viscum album* is parasitic on more than 230 hosts, including over 180 introduced species; in southern Europe it causes much damage to forest crops, and has also intermittently been regarded as a pest on fruit trees in the west country. It is most common on apple trees. Despite its reputation, it is rare to find it growing on oak; Elwes and Henry, after an exhaustive search, could list only twenty-three reported cases in their *Trees of Great Britain and Ireland* (1907). Indeed, it was probably this rarity that caused the Druids to prize oak-colonizing specimens for use in their rituals. In the nineteenth century, mistletoe was sometimes grafted onto different trees for ornamental effect, and in 1837 Donald Beaton published instructions for this process in the *Gardener's Magazine*.

Watercolour on paper, 57 × 39 cm
PROVENANCE: Presented by the artist's sister, 1949

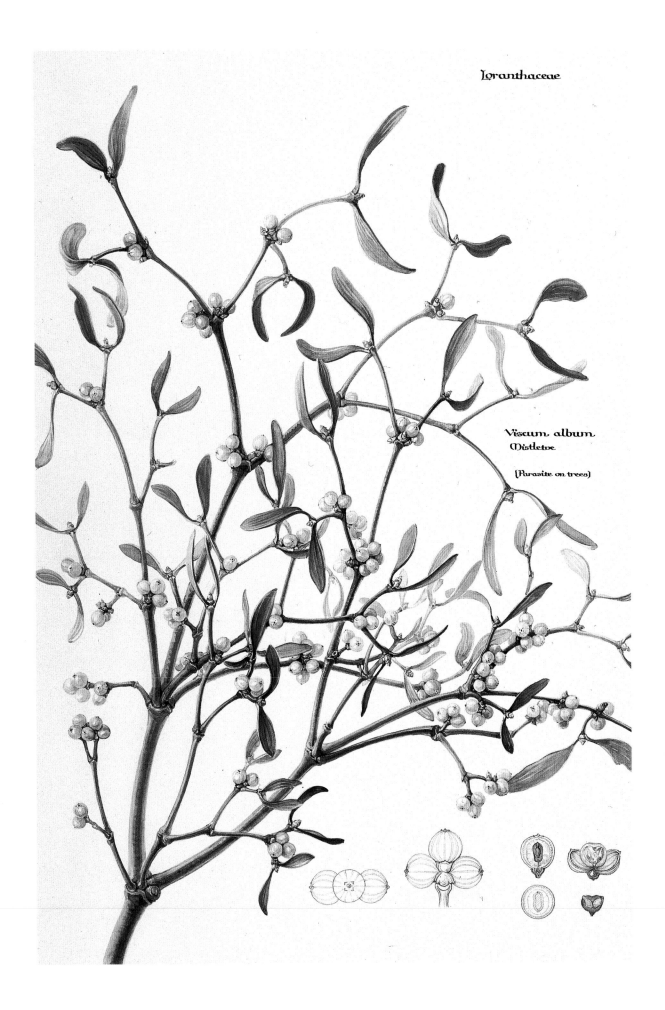

Loranthaceae

Viscum album
Mistletoe

(Parasite on trees)

PLATE 59

Vera Higgins (1892–1968)
Hibiscus syriacus, unnamed form

Malvaceae

Vera Higgins, née Cockburn, was born in 1892 and studied at Swanley College. Most of her publications, whether translations or her own work, were on succulents, among them *The Study of Cacti* (1933), *Succulents in Cultivation* (1960), and *Crassulas in Cultivation* (1964). She edited the *Cactus Journal* from 1932 to 1939, the *Alpine Garden Society Bulletin* from 1938 to 1947, and the *Journal of the Royal Horticultural Society* from 1941 to 1945. She was a member of the Society's Picture Committee, and exhibited her drawings on several occasions at shows. She was awarded the Victoria Medal in 1946. Vera Higgins bequeathed her library and papers to the Society; part of this bequest consisted of 693 drawings, mostly of orchids and succulents.

The tree mallow, *Hibiscus syriacus*, is first mentioned in England by John Gerard, who in 1597 reported that he had sown seeds of it and was 'expecting the success'. Despite its name, it is native to India and China, and consequently had difficulties in the English climate during the seventeenth century, when gardeners described it as more tender than it would be considered today. Nonetheless, Philip Miller could list seven sorts by the 1750s, and their number gradually increased over the next two centuries.

Higgins' dissection of the pistil, on the upper left of the drawing, shows the influence of the illustrations in A. H. Church's *Types of Floral Mechanism* (1908).

Watercolour on paper, 29.5 × 23 cm
PROVENANCE: Vera Higgins bequest, 1968

Hibiscus ?

Malvaceae

July 1928

PLATE 60

Margaret Stones (1920–   )
Lilium candidum   Liliaceae

E. Margaret Stones was born in Victoria, Australia, in 1920. She came to England in 1951, where she worked at Kew and the British Museum (Natural History). In 1955 the Society hired her to succeed Lilian Snelling as artist for *Curtis's Botanical Magazine*, a position she held until the title ceased in 1983. While working for the Society, she also made illustrations for F. C. Stern's *Snowdrops and Snowflakes* (1956) and the last two *Supplements to Elwes' Monograph of the Genus Lilium* (1960–62). Her major work may well be the illustrations for Winifred Curtis's *Endemic Flora of Tasmania* (six volumes, 1967–78).

Lilium candidum is native to the eastern Mediterranean and Balkans. It was commonly known as the white lily until the end of the eighteenth century, when other white lilies began to be imported into England. In the nineteenth century, when there was much interest in recovering (or, if need be, inventing) the traditional folklore of plants, it became known instead as the 'Madonna lily', since it was often depicted in paintings of the Virgin Mary. Patrick Synge, in his *Lilies* (1980), said: 'Undoubtedly our ancestors grew this lily better than we do today', because of the prevalence of disease in recent times.

For this drawing, made in 1968 for Patrick Synge's *Lilies*, a revision of H. J. Elwes' famous monograph, Miss Stones painted the flowers from a specimen grown by Mrs F. Dundas of Ryarsh, Kent, and the bulb from a specimen at Kew.

Watercolour on paper, 40.5 × 30 cm
PROVENANCE: Commissioned 1968 for use in Patrick Synge's *Lilies*

PLATE 61

Margaret Stones (1920–   )
Cardiocrinum giganteum var *yunnanense*   Liliaceae

The giant lily of the Himalayas was first offered for sale in England by the Veitch nurseries in Chelsea in 1851. Although John Lindley had proposed making *Cardiocrinum* a separate genus in 1847, it was not until just over a century later that his suggestion was generally adopted, following W. T. Stearn's revision of *Lilium*. 'There can hardly be a more exciting moment for a Lily enthusiast', wrote Stearn, 'than to see this noble Lily in its native haunts for the first time.'

The naturally occurring variety *yunnanense* is distinguished from the species because the flowers open from the top of the stem down. E. H. Wilson sent bulbs to the Veitch nurseries, and seeds to America, at the turn of the century, without success; it was introduced into cultivation by the Baden nurseryman Max Leichtlin, who obtained it from French priests in China. In 1968 a spike of this plant from Windsor Great Park received an Award of Merit, when exhibited by the Crown Estate Commissioners; this drawing, made from a specimen at Windsor, was reproduced in Synge's *Lilies* (1980). Oliver Wyatt, in an RHS lecture for young people in 1967, warned: 'you may be disheartened by this because, from seed, it takes ten years and when it's flowered, it dies'.

Watercolour on paper, 56 × 38 cm
PROVENANCE: Commissioned 1968 for use in Patrick Synge's *Lilies*

PLATE 62

John Paul Wellington Furse (1904–1976)
Narcissus species and varieties   Amaryllidaceae

John Paul Wellington Furse was born in 1904, and entered the Navy at the age of thirteen. The minute books of the RHS Picture Committee record his gradual ascent in rank, from Lieutenant in the 1930s, when he first began exhibiting pictures at RHS shows, to Rear-Admiral. In 1959 he retired from the Navy, and began a second career as a plant collector, travelling first with Patrick Synge in 1960, and on later occasions with his wife, in Iran, Afghanistan, and the Near East. He died in 1976, leaving a large quantity of drawings at Kew and the Lindley Library.

This drawing, made in April 1951, records seven types of daffodil. From left to right, these are: *Narcissus cyclamineus*, *N. bulbocodium*, *N. rugulosus* 'Orange Queen', *N. triandrus albus*, *N. jonquilla*, *N. rugulosus*, and *N. pseudonarcissus*. 'Orange Queen' is a sport of *N. rugulosus*, which was given an Award of Merit in Haarlem in 1913; since the firm of Cartwright and Goodwin introduced a quite different 'Orange Queen' in 1908, the two forms became confused in commerce. The purpose of plant registration is to avoid confusions like this.

Watercolour on paper, 35.5 × 25.5 cm
PROVENANCE: Bequest from the artist, 1976

Eremocarpus scaber

PLATE 63

Mary Grierson (1912–   )
Eccremocarpus scaber   Bignoniaceae

In the nineteenth century, Chelsea was the site of some of the country's most famous nurseries, trading largely in exotic plants to be grown under glass. From the 1840s on, the Veitch nurseries were probably the most famous, but in the 1820s and 1830s one of the principal nurserymen of the area was James Charles Tate of the Sloane Street Nursery. Among the plants sent to him by Robert Ponsonby Staples, the British consul in Latin America, was the Chilean glory flower, *Eccremocarpus scaber*, which he first raised in 1824. By the end of the century, the ranks of the Chelsea nurseries were thinning, as London air pollution made cultivation ever more difficult; William Bull and Sons were the last to leave, at the end of World War I.

Mary Grierson studied botanical art under John Nash. She was appointed Botanical Illustrator at Kew in 1960, and has contributed many plates to *Curtis's Botanical Magazine* and its successor, *The Kew Magazine*. She has undertaken commissions for the World Wide Fund for Nature, the Israeli Nature Reserve Authority, and the Pacific Tropical Botanical Garden in Hawaii. She has received five Gold Medals at Royal Horticultural Society shows, and in 1985 was awarded the Gold Veitch Medal. Her most important published works have been *Orchidaceae* (1973) and *The Country Life Book of Orchids* (1978), with texts by Peter Hunt, and *An English Florilegium: the Tradescant Legacy* (1987), with text by C.D.Brickell. She is a member of the Society's Picture Committee.

Ink and watercolour on paper, 40.5 × 28 cm
PROVENANCE: Purchased 1993

PLATE 64

Jenny Brasier (1936–   )
Parrotia persica   Hamamelidaceae

Parrotia was named in honour of Johann Jacob Parrot, who in 1829 became the first man to climb Mount Ararat. In 1868, describing *Parrotia persica* in *Curtis's Botanical Magazine*, Sir Joseph Hooker could call it one of the rarest trees in cultivation: 'Two small trees of it exist in the Royal Gardens, which were received as pot-plants from St Petersburg twenty-five to thirty years ago'. In 1884, when William Paul, of the Royal Nurseries at Waltham Cross, exhibited it, it received a First Class Certificate. In 1891 Sir Harry Veitch described it as 'one of the handsomest of shrubs for its autumnal foliage', and with the increasing interest in

autumn colour during the twentieth century, it became more widespread in English gardens.

Jenny Brasier, of Merrist Wood, has won four Gold Medals at RHS shows. The drawing illustrated here is one of two drawings of leaves, made for the late Wilfrid Blunt in 1987, at a time when he was considering revising his *Art of Botanical Illustration*, and which were purchased by the Society from her Gold Medal-winning exhibit in February 1988.

Watercolour on vellum, 13.5 × 11 cm
PROVENANCE: Purchased 1988

PLATE 65

Raymond Booth (1929–   )
Lathyrus odoratus   Leguminosae

Interest in sweet pea breeding slowly grew during the nineteenth century; but it was not until the last quarter of the century that Henry Eckford, of Wem in Shropshire, made it truly a popular favourite. In 1900, an exhibition was held to commemorate the bicentenary of the sweet pea's introduction, and in its wake the National Sweet Pea Society was founded. The following year the first waved sweet pea was introduced by Silas Cole, head gardener at Althorp, under the name 'Countess Spencer', and the sweet pea craze began. In 1912, the *Daily Mail* sponsored a sweet pea competition with a £1000 first prize. Walter P. Wright, the founder of the magazine *Popular Gardening*, could write in his book *The Perfect Garden* (1908): 'I dream of a sweet-pea garden. This has no old-time flavour ... It is modern, strenuous, fiercely vital. The flower is in the fire of transformation by the florist, and new varieties pour out hotly, like the editions of evening newspapers.' Between the wars, some four or five hundred cultivars were available; since those heady days, the number has declined.

Raymond C. Booth was born in Leeds in 1929. In the early 1950s his work attracted attention at the Royal Horticultural Society's shows for his technique of painting plants in oils; in 1953, the Society purchased some of his paintings, and commissioned others for use as illustrations for the *RHS Journal* and *Some Good Garden Plants*. He collaborated with Paul Jones on the illustrations for Beryl Urquhart's *Camellia* (1956). His most recent work is a book about Japanese plants entitled *Japonica Magnifica* (1992).

Oil on card, 99 × 65 cm
PROVENANCE: Purchased, 1991

148

PLATE 66

Jenny Jowett (1936–)
Helleborus × sternii Blackthorn Strain   Ranunculaceae

The sort of popularity that sweet peas once enjoyed has now fallen to hostas and, increasingly, hellebores.

In the 1940s, Sir Frederick Stern raised some new hybrid hellebores at his garden at Highdown, near Worthing in Sussex. The result of crossing *H. argutifolius* and *H. lividus*, they were named *Helleborus × sternii*, and first introduced into commerce by the firm of Hilling in 1947.

The hybrid Blackthorn Strain was raised by Robin White of the Blackthorn Nursery, Alresford, Hampshire, by crossing *H. × sternii* with cultivars of *H. nigra*; Peter Chappell of Spinners was the first nurseryman to retail it in the late 1980s. It is characterized by its compact growth, purple stems, grey-green leaves, and slightly pink-flushed green flowers.

Jenny Jowett, of Silchester, has exhibited at RHS shows since the mid 1970s, and has won two Gold Medals. On the second of these occasions, in 1989, this coloured drawing of *Helleborus × sternii* Blackthorn Strain was purchased for the Society's collection. Her illustrations for Diana Grenfell's *The White Garden* (1990) received a Silver-Gilt Grenfell Medal.

Watercolour on paper, 46 × 35 cm
PROVENANCE: Purchased 1989

Helleborus x sternii Blackthorn strain   Jenny Jowett

PLATE 67

Charles Stitt (1941–)
Apple (*Malus domestica*) 'Falstaff'   Rosaceae
Pear (*Pyrus communis*) 'Concorde'

The programme of fruit breeding inaugurated in the early nineteenth century by Thomas Andrew Knight continues today, as shown in this illustration of two new varieties illustrated in *The Plantsman* in 1990.

The apple 'Falstaff' is the result of a cross between 'James Grieve' and 'Golden Delicious'. 'James Grieve', named after its raiser, was introduced by Dicksons of Edinburgh in the 1890s. 'Golden Delicious' was raised in America about 1890, and introduced by Stark Brothers in 1914; it was introduced into England by E. A. Bunyard in his 1927/8 catalogue, and has since become one of the country's dominant commercial apples. 'Falstaff' germinated in 1966, and was sent to the National Fruit Trials at Brogdale in 1973; it crops twice as heavily as 'Cox's Orange Pippin'.

'Concorde' is a cross between two of the best-known pears, 'Doyenné du Comice' (familiarly known to many as Comice), and 'Conference'. 'Doyenné du Comice' first fruited in 1849, in the garden of the Comice Horticole in Angers, and was introduced into England by Sir Thomas Dyke Acland of Killerton, Devon, in 1858. 'Conference', bred by Rivers' nursery in Sawbridgeworth, Hertfordshire, took its name from the Royal Horticultural Society's Pear Conference in 1885, where it was first exhibited; it was introduced into commerce in 1894. 'Concorde' germinated in 1968, and was sent for trial at Brogdale in 1979.

Charles Stitt is best known for his illustrations for Will Ingwersen's *Classic Garden Plants* (1975). In addition, he has contributed colour plates and numerous line drawings for *The Plantsman*.

Watercolour on paper, 43 × 30 cm
PROVENANCE: Purchased 1992

PLATE 68

Pauline Dean (1943–)
Iris tectorum, Burma form   Iridaceae

Iris tectorum has been cultivated since the seventh century in China, where its root was used to make a powder for whitening the skin. It was introduced early on into Japan, where it was grown on the ridges of thatched roofs. Engelbert Kaempfer, a physician for the Dutch East India Company, described it in 1712 under the name 'Itz fatz'. It was introduced into Europe by Franz von Siebold, a German botanist who travelled in Japan at a time when it was still officially closed to the West, and who, on his return to Europe, established an acclimatization garden and nursery for Japanese plants. It was originally called *Iris chinensis*, but as that name had already been used previously, it is now known by a specific epithet coined by Maximowicz in 1871. Its cultivation in English gardens was delayed during the nineteenth century by doubts about its hardiness, but after Reginald Farrer pointed out how well it did in an exposed site in Yorkshire, its popularity increased.

Pauline Dean, of Guildford, received her first Gold Medal from the Society in 1989. In 1991 the drawing shown here was purchased for the Society. She is one of the illustrators of the *New RHS Dictionary of Gardening*, and teaches botanical illustration at the Society's garden at Wisley.

Watercolour on paper, 47 × 35 cm
PROVENANCE: Purchased 1991

PLATE 69

Ann Farrer (1950–   )
Ilex colchica   Aquifoliaceae

The first exotic holly was introduced into England in 1744, to be followed by many other species from America and China over the next two centuries. The nineteenth century saw the beginnings of holly hybridization. The most important early hybrid was *Ilex × altaclerensis*, raised by J. R. Gowen, later the Secretary of the Horticultural Society, in his garden at Highclere (of which *altaclerensis* is a Latinization) in the 1830s; this has been the parent of a wide range of modern cultivars, which share the trait of smooth-edged leaves.

New hollies continue to be identified today, and the species illustrated here, *Ilex colchica*, was first described in 1947 by the Russian botanist Antonina Ivanovna Pojarkova (1897–1980). Native to Turkey and the Caucasus, it differs from *Ilex aquifolium* in having less undulating leaves and forward-pointing spines. It was introduced into England by Roy Lancaster, about 1979; the specimen from which this drawing was made was brought into this country by Martyn Rix, in 1981.

Ann Farrer first exhibited drawings at an RHS show in 1982, when she received the first of her five Gold Medals to date. Formerly married to the plant collector Tony Schilling, she has made many drawings of Himalayan plants, and illustrated Polunin and Stainton, *Flowers of the Himalaya* (1984). Another drawing of hers in the possession of the Society, *Pinus wallichiana*, was reproduced in W. T. Stearn's *Flower Artists of Kew* (1990), published by the Herbert Press. In 1988 the Linnean Society made her the first recipient of their Jill Smythies Award.

Watercolour on paper, 39 × 31 cm
PROVENANCE: Commissioned for *The New Plantsman* and purchased 1993

PLATE 70

J. L. Macfarlane (1838–*c.* 1913)
Nellie Roberts (1872–1959)
M. Iris Humphreys (*fl.* 1930s–1970s)
Cherry-Anne Lavrih (1946–   )
Four orchid award portraits   Orchidaceae

J. L. Macfarlane pioneered the style and format that have since been generally adopted for portraying orchid cultivars: frontal view of the flower, natural size, sometimes with tinted background for emphasis, especially if parts of the flower are white. The drawing shown here depicts an orchid grown by Sir Trevor Lawrence, President of the Royal Horticultural Society from 1883 to 1913; Sir Trevor's son, Sir William Lawrence, presented a collection of Macfarlane's drawings to Gurney Wilson, who described them in the *Orchid Review*, and in turn bequeathed them to the Society.

The style that Macfarlane developed for portraying orchid cultivars was copied by other artists, sometimes even to his style of lettering. Nellie Roberts, the first orchid painter hired by the Royal Horticultural Society, based her work on his; for over sixty years, from 1897 to her death, she painted portraits of orchid cultivars that had been given awards.

Of the artists who succeeded her, two are represented in this plate. M. I. Humphreys began her career in the 1930s, painting portraits of the orchid hybrids raised by the firm of Charlesworth, and then moved to Armstrong and Brown, of which her husband became a director; she worked for the Society from 1967 to 1980, and received the Westonbirt Orchid Medal in 1977 for her work. Cherry-Anne Lavrih, formerly head of the Art Department at Whyteleafe Grammar School for Girls, has been the Society's orchid painter since 1987.

Orchids portrayed: top left (Macfarlane), *Paphiopedilum* 'Euryades', 1898; top right (Roberts), *Brassolaeliocattleya* 'Fowleri', 1907; bottom left (Humphreys), *Cymbidium* Magna Charta 'Spring Promise', 1968; bottom right (Lavrih), *Dracula chiroptera* 'Pelleas', 1992.

Watercolour on board, average size 28 × 22 cm
PROVENANCE: Macfarlane: Gurney Wilson bequest, 1957. The others commissioned by the Royal Horticultural Society at various dates.

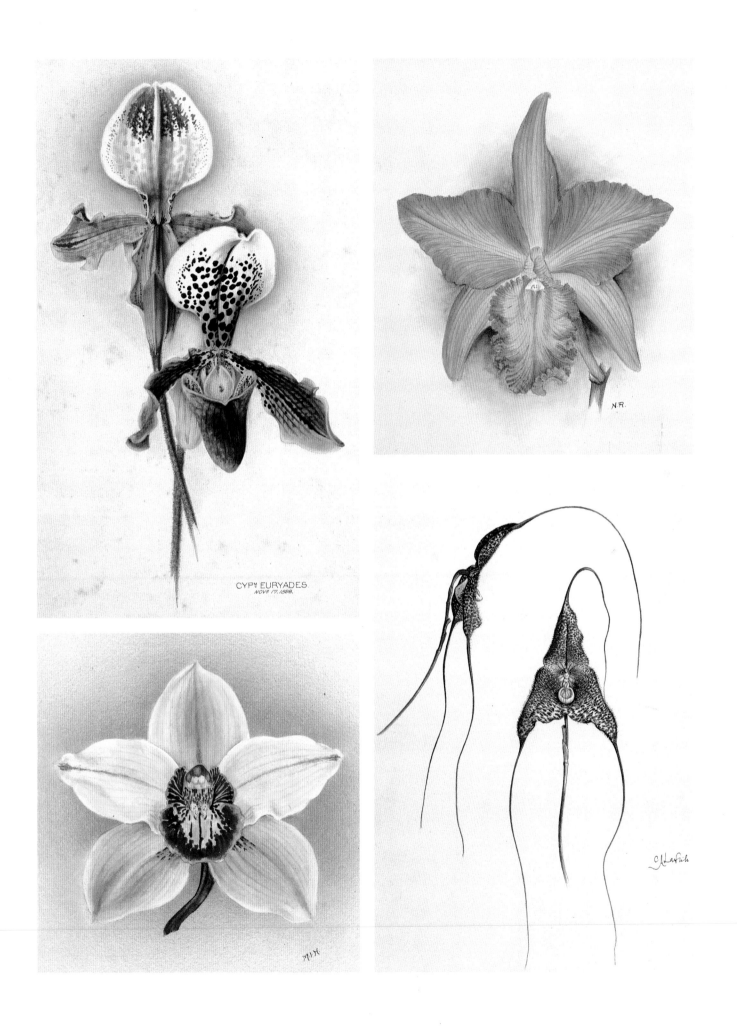

CYP^m EURYADES.
NOV⁸ 17, 1898.

Index of Plant Names

Index of Artists